Foseco Non-Ferrous Foundryman's Handbook

Foseco Non-Ferrous Foundryman's Handbook

Eleventh edition

Revised and edited by
John R. Brown

Acknowledgement
The author gratefully acknowledges the permission
to reproduce the article represented by chapter 7 which
was written by Jeff Meredith, Casting Solutions Pty Ltd.

ELSEVIER
BUTTERWORTH
HEINEMANN

AMSTERDAM BOSTON HEIDELBERG LONDON NEW YORK OXFORD
PARIS SAN DIEGO SAN FRANCISCO SINGAPORE SYDNEY TOKYO

Elsevier Butterworth-Heinemann
Linacre House, Jordan Hill, Oxford OX2 8DP
30 Corporate Drive, Burlington, MA 01803

Ninth edition published by Pergamon Press plc 1986
Tenth edition 1994
Reprinted 1995, 1996, 1998
Eleventh edition 1999
Reprinted 2002, 2004, 2005

British Library Cataloguing in Publication Data
Brown, John R.
  Foseco non-ferrous foundryman's handbook
  1. Nonferrous metals – Founding – Handbooks, manuals, etc.
  I. Title
  673

Library of Congress Cataloguing in Publication Data
Foseco non-ferrous foundryman's handbook/revised and edited by John R. Brown –
11th ed.
p. cm.
Prev. eds. Published under title: Foseco foundryman's handbook.
1. Founding Handbooks, manuals, etc. 2. Nonferrous metals –
Founding Handbooks, manuals, etc. I. Brown, John R. III. Title:
Non-ferrous foundryman's handbook   III. Title: Foseco foundryman's handbook.
TA235.F585   99-34985
673-dc21   CIP

ISBN 0 7506 4286 6

For information on all Elsevier Butterworth-Heinemann
publications visit our website at www.bh.com

Transferred to digital print 2008
Printed and bound by CPI Antony Rowe, Eastbourne

**Working together to grow
libraries in developing countries**

www.elsevier.com | www.bookaid.org | www.sabre.org

ELSEVIER   BOOK AID International   Sabre Foundation

Composition by Genesis Typesetting, Rochester, Kent

Contents

Chapter 4 Fluxes

Preface

The last edition of the *Handbook* was published in 1994 and like all the earlier editions, it aimed to provide a practical reference book for all those involved in making castings in any of the commonly used alloys by any of the usual moulding methods. In order to keep the *Handbook* to a reasonable size, it was not possible to deal with all the common casting alloys in detail. Since 1994 the technology of casting has continued to develop and has become more specialised so that it has been decided to publish the 11th edition of the *Handbook* in three separate volumes:

Non-ferrous	dealing with aluminium, copper and magnesium casting alloys
Iron	dealing with grey, ductile and special purpose cast irons
Steel	dealing with carbon, low alloy and high alloy steels

Certain chapters (with slight modifications) are common to all three volumes: these chapters include tables and general data, sands and sand bonding systems, resin bonded sand, sodium silicate bonded sand and feeding systems. The remaining chapters have been written specifically for each volume.

The *Handbook* refers to many Foseco products. Not all of the products are available in every country and in a few cases, product names may vary. Users should always contact their local Foseco company to check whether a particular product or its equivalent is available.

The Foseco logo and all product names appearing in capital letters are trademarks of the Foseco group of companies, used under licence.

John R. Brown

Acknowledgements

The following Organisations have generously permitted the use of their material in the *Handbook*:

The American Foundrymen's Society, Inc., 505 State Street, Des Plaines, Illinois 60016-8399, USA.

The Association of Light Alloy Founders (ALARS), Broadway House, Calthorpe Road, Five Ways, Birmingham, B15 1TN.

BSI, Extracts from British Standards are reproduced with the permission of British Standards Institution. Complete copies can be obtained by post from Customer Services, BSI, 389 Chiswick High Road, London W4 4AL.

Buhler UK Ltd, 19 Station Road, New Barnet, Herts, EN5 1NN.

Butterworth-Heinemann, Linacre House, Jordan Hill, Oxford OX2 8DP.

The Castings Development Centre (incorporating BCIRA), Bordesley Hall, The Holloway, Alvechurch, Birmingham, B48 7QB.

The Castings Development Centre (incorporating Steel Castings Research & Trade Association), 7 East Bank Road, Sheffield, S2 3PT.

Chem-Trend (UK) Ltd, Bromley Street, Lye, Stourbridge, West Midlands DY9 8HY.

Copper Development Association, Verulam Industrial Estate, 224, London Road, St. Albans, Herts, AL1 1AQ.

Foundry International, DMG Business Media Ltd, Queensway House, 2 Queensway, Redhill, Surrey, RH1 1QS.

Foundry Management & Technology, 1100 Superior Avenue, Cleveland, OH 44114, USA.

Foundry & Technical Liaison Ltd, 6-11 Riley Street, Willenhall, West Midlands, WV13 1RH.

The Institute of British Foundrymen, Bordesley Hall, The Holloway, Alvechurch, Birmingham, B48 7QA.

International Magnesium Association, 1303 Vincent Place, Suite One, McLean, Virginia 22101, USA.

OEA (Organisation of European Aluminium Refiners and Remelters, Broadway House, Calthorpe Road, Five Ways, Birmingham, B15 1TN.

Ramsell Furnaces Ltd, Wassage Way, Hampton Lovett Industrial Estate, Droitwich, Worcestershire, WR9 0NX.

Striko UK Ltd, Newcastle Street, Stone, Staffordshire, ST15 8JT.

The author gratefully acknowledges the help received from many individuals, in particular from colleagues at Foseco.

All statements, information and data contained herein are published as a guide and although believed to be accurate and reliable (having regard to the manufacturer's practical experience) neither the manufacturer, licensor, seller nor publisher represents or warrants, expressly or implied:

1 Their accuracy/reliability
2 The use of the product(s) will not infringe third party rights
3 No further safety measures are required to meet local legislation.

The seller is not authorised to make representations nor contract on behalf of the manufacturer/licensor. All sales by the manufacturer/seller are based on their respective conditions of sale available on request.

Chapter 1

Tables and general data

SI units and their relation to other units

The International System of Units (SI System) is based on six primary units:

Quantity	Unit	Symbol
length	metre	m
mass	kilogram	kg
time	second	s
electric current	ampere	A
temperature	degree Kelvin	K
luminous intensity	candela	cd

Multiples

SI prefixes are used to indicate multiples and submultiples such as 10^6 or 10^{-3}

	Prefix	Symbol		Prefix	Symbol
10	deca	da	10^{-1}	deci	d
10^2	hecto	h	10^{-2}	centi	c
10^3	kilo	k	10^{-3}	milli	m
10^6	mega	M	10^{-6}	micro	μ
10^9	giga	G	10^{-9}	nano	n
10^{12}	tera	T	10^{-12}	pico	p

Example: One millionth of a metre is expressed as one micrometre, 1 μm.

Derived units

The most important derived units for the foundryman are:

Quantity	Unit	Symbol
Force	newton	N ($kg\ m/s^2$)
Pressure, stress	newton per square metre or pascal	N/m^2 (Pa)
Work, energy	joule	J (Nm)
Power, heat flow rate	watt, joule per second	W (J/s)
Temperature	degree Celsius	°C
Heat flow rate	watt per square metre	W/m^2
Thermal conductivity	watt per metre degree	W/m K
Specific heat capacity	joule per kilogram degree	J/kg K
Specific latent heat	joule per kilogram	J/kg

SI, metric, non-SI and non-metric conversions

Length:
1 in = 25.4 mm
1 ft = 0.3048 m
1 m = 1.09361 yd
1 km = 1093.61 yd = 0.621371 miles
1 mile = 1.60934 km = 1760 yd
1 yd = 0.9144 m

Area:
$1\ in^2$ = 654.16 mm^2
$1\ ft^2$ = 0.092903 m^2
$1\ m^2$ = 1.19599 yd^2 = 10.76391 ft^2
$1\ mm^2$ = 0.00155 in^2
$1\ yd^2$ = 0.836127 m^2
1 acre = 4840 yd^2 = 4046.86 m^2 = 0.404686 m^2 hectare
1 hectare = 2.47105 acre = 10 000 m^2

Volume:
$1\ cm^3$ = 0.061024 in^3
$1\ dm^3$ = 1 l (litre) = 0.035315 ft^3
$1\ ft^3$ = 0.028317 m^3 = 6.22883 gal (imp)
1 gal (imp) = 4.54609 l (litre)
$1\ in^3$ = 16.3871 cm^3
1 l (litre) = 1 dm^3 = 0.001 m^3 = 0.21997 gal (imp)
$1\ m^3$ = 1.30795 yd^3 = 35.31467 ft^3

1 pt (pint)   = 0.568261 l
1 US gal   = 3.785411 = 0.832674 gal (imp)
1 ft^3/min (cfm) = 1.699 m^3/h
1 ft^3/sec   = 28.31681 l/s

Mass:
1 lb (pound) = 0.453592 kg
1 cwt   = 50.802 kg
1 kg   = 2.20462 lb
1 oz   = 28.349 gm
1 ton   = 2240 lb = 1.01605 t (tonne) = 1016.05 kg
1 ton (US)   = 2000 lb = 907.185 kg

Force:
1 kgf   = 9.80665 N = 2.20462 lbf = 1 kp (kilopond)
1 lbf   = 4.44822 N
1 pdl (poundal) = 0.138255 N

Density:
1 kgf/m^3 = 0.062428 lb/ft^3
1 lb/ft^3 = 16.0185 kg/m^3
1 g/cm^3 = 1000 kg/m^3

Pressure, stress:
1 kgf/cm^2 = 98.0665 kPa (kN/m^2)
1 kgf/mm^2 = 9.80665 N/mm^2 = 1422.33 lbf/in^2 = 0.63497 tonf/in^2
1 lbf/in^2 (psi)   = 6.89476 kPa (kN/m^2)
1 Pa (N/m^2)   = 0.000145038 lbf/in^2
1 in w.g. (in H$_2$O) = 249.089 Pa
1 N/mm^2 = 1 MPa = 145.038 lbf/in^2 = 0.06475 tonf/in^2
                    = 0.10197 kgf/cm^2

Power:
1 kW   = 3412 Btu/hr
1 hp (horsepower) = 0.745700 kW

Energy, heat, work:
1 Btu   = 1.05506 kJ
1 cal   = 4.1868 J
1 kWh   = 3.6 MJ = 3412 Btu
1 therm = 100 000 Btu = 105.506 MJ
1 kJ   = 0.277778 W.h

Specific heat capacity, heat transfer:
1 cal/g°C   = 1 kcal/kg°C = 4186.8 J/kg.K
1 Btu/lb°F   = 4186.8 J/kg.K
1 Btu/h   = 0.293071 W

1 cal/cm.s°C   = 418.68 W/m.K (thermal conductivity)
1 Btu.in/ft^{2}h°F = 0.144228 W/m.K (thermal conductivity)
1 Btu/ft^{2}h°F   = 5.67826 W/m^2.K (heat transfer coeff.)

Miscellaneous:
1 std.atmos.   = 101.325 kPa = 760 mm Hg = 1.01325 bar
1 bar   = 100 kPa = 14.5038 lbf/in^2
1 cP (centipoise) = 1 mPa.s
1 cSt (centistoke) = 1 mm^2/s
1 cycle/s   = 1 Hz (Hertz)
1 hp   = 745.7 W

Useful approximations:

1 Btu	= 1 kJ		1 kg	= 2¼ lb
1 ft	= 30 cm		1 kgf	= 10 N
1 gal	= 4½ l		1 std atmos.	= 1 bar
1 ha	= 2½ acre		1 km	= ⅝ mile
1 hp	= ¾ kW		1 litre	= 1¾ pint
1 in	= 25 mm		1 lbf	= 4½ N
1 therm	= 100 MJ		1 yd	= 0.9 m
1 tonf/in^2	= 15 N/mm^2			
1 psi (lbf/in^2)	= 7 kPa			
1 N (newton)	= the weight of a small apple!			

Temperature:
°F   = 1.8 × °C + 32
°C   = (°F − 32)/1.8
0°C (Celsius) = 273.15 K (Kelvin)

Conversion table of stress values

American (lb/in^2)	British (ton/in^2)	Metric (kgf/mm^2)	SI (N/mm^2)
		Equivalent stresses	
250	0.112	0.176	1.724
500	0.223	0.352	3.447
1000	0.446	0.703	6.895
2000	0.893	1.406	13.789
3000	1.339	2.109	20.684
4000	1.788	2.812	27.579
5000	2.232	3.515	34.474
10 000	4.464	7.031	68.947
15 000	6.696	10.546	103.421
20 000	8.929	14.062	137.894
25 000	11.161	17.577	172.368
30 000	13.393	21.092	206.841
35 000	15.652	24.608	241.315
40 000	17.875	28.123	275.788
45 000	20.089	31.639	310.262
50 000	22.321	35.154	344.735
55 000	24.554	38.670	379.209
60 000	26.786	42.185	413.682
65 000	29.018	45.700	448.156
70 000	31.250	49.216	482.629
75 000	33.482	52.731	517.103
80 000	35.714	56.247	551.576
85 000	37.946	59.762	586.050
90 000	40.179	63.277	620.523
95 000	42.411	66.793	654.997
100 000	44.643	70.308	689.470

Conversions

10 000	4.464	7.031	68.947
22 399	**10**	15.749	154.438
14 223	6.349	**10**	98.066
14 504	6.475	10.197	**100**

Areas and volumes of circles, spheres, cylinders etc.

π = 3.14159 (approximation: 22/7)

1 radian = 57.296 degrees

Circle; radius r, diameter d:

circumference = $2\pi r = \pi d$

area = $\pi r^2 = \pi/4 \times d^2$

Sphere; radius r:

surface area = $4\pi r^2$

volume = $\frac{4}{3}\pi r^3$

Cylinder; radius of base r, height h:

area of curved surface = $2\pi rh$

volume = $\pi r^2 h$

Cone; radius of base r, height h:

volume = $\frac{1}{2}$ area of base $\times$ height

= $\frac{1}{2}\pi r^2 h$

Triangle; base b, height h:

area = $\frac{1}{2}bh$

The physical properties of metals

Element	Symbol	Atomic weight	Melting point (°C)	Boiling point (°C)	Latent heat of fusion (kJ/kg)	(cal/g)	Mean specific heat 0–100°C (kJ/kg.K)	(cal/g°C)
Aluminium	Al	26.97	660.4	2520	386.8	92.4	0.917	0.219
Antimony	Sb	121.76	630.7	1590	101.7	24.3	0.209	0.050
Arsenic	As	74.93	volat.	616	–	–	0.331	0.079
Barium	Ba	137.37	729	2130	–	–	0.285	0.068
Beryllium	Be	9.02	1287	2470	133.5	31.9	2.052	0.490
Bismuth	Bi	209.0	271.4	1564	54.4	13.0	0.125	0.030
Cadmium	Cd	112.41	321.1	767	58.6	14.0	0.233	0.056
Calcium	Ca	40.08	839	1484	328.6	78.5	0.624	0.149
Carbon	C	12.01	–	–	–	–	0.703	0.168
Cerium	Ce	140.13	798	3430	–	–	0.188	0.045
Chromium	Cr	52.01	1860	2680	132.7	31.7	0.461	0.110
Cobalt	Co	58.94	1494	2930	244.5	58.4	0.427	0.102
Copper	Cu	63.57	1085	2560	180.0	43.0	0.386	0.092
Gallium	Ga	69.74	29.7	2205	80.2	19.2	0.377	0.090
Gold	Au	197.2	1064.4	2860	67.4	16.1	0.130	0.031
Indium	In	114.8	156	2070	–	–	0.243	0.058
Iridium	Ir	193.1	2447	4390	–	–	0.131	0.031
Iron	Fe	55.84	1536	2860	200.5	47.9	0.456	0.109
Lead	Pb	207.22	327.5	1750	20.9	5.0	0.130	0.031
Lithium	Li	6.94	181	1342	137.4	32.8	3.517	0.840
Magnesium	Mg	24.32	649	1090	194.7	46.5	1.038	0.248
Manganese	Mn	54.93	1244	2060	152.8	36.5	0.486	0.116
Mercury	Hg	200.61	–38.9	357	12.6	3.0	0.138	0.033
Molybdenum	Mo	96.0	2615	4610	–	–	0.251	0.060
Nickel	Ni	58.69	1455	2915	305.6	73.0	0.452	0.108
Niobium	Nb	92.91	2467	4740	–	–	0.268	0.064
Osmium	Os	190.9	3030	5000	–	–	0.130	0.031
Palladium	Pd	106.7	1554	2960	150.7	36.0	0.247	0.059
Phosphorus	P	31.04	44.1	279	20.9	5.0	0.791	0.189
Platinum	Pt	195.23	1770	3830	113.0	27.0	0.134	0.032
Potassium	K	39.1	63.2	759	67.0	16.0	0.754	0.180
Rhodium	Rh	102.91	1966	3700	–	–	0.243	0.058
Silicon	Si	28.3	1412	3270	502.4	120.0	0.729	0.174
Silver	Ag	107.88	961.9	2163	92.1	22.0	0.234	0.055
Sodium	Na	23.00	97.8	883	115.1	27.5	1.227	0.293
Strontium	Sr	87.63	770	1375	–	–	0.737	0.176
Sulphur	S	32.0	115	444.5	32.7	9.0	0.068	0.016
Tantalum	Ta	180.8	2980	5370	154.9	37.0	0.142	0.034
Tellurium	Te	127.6	450	988	31.0	7.4	0.134	0.032
Thallium	Tl	204	304	1473	–	–	0.130	0.031
Tin	Sn	118.7	232	2625	61.1	14.6	0.226	0.054
Titanium	Ti	47.9	1667	3285	376.8	90.0	0.528	0.126
Tungsten	W	184.0	3387	5555	167.5	40.0	0.138	0.033
Uranium	U	238.2	1132	4400	–	–	0.117	0.029
Vanadium	V	50.95	1902	3410	334.9	80.0	0.498	0.119
Zinc	Zn	65.38	419.6	911	110.1	26.3	0.394	0.094
Zirconium	Zr	90.6	1852	4400	–	–	0.289	0.069

The physical properties of metals (*Continued*)

Element	Thermal conductivity (W/m.K)	Resistivity (μohm.cm at 20°C)	Vol. change on melting (%)	Density (g/cm^3)	Coeff. of expansion ($\times 10^{-6}$/K)	Brinell hardness no.
Al	238	2.67	6.6	2.70	23.5	17
Sb	23.8	40.1	1.4	6.68	11	30
As	–	33.3	–	5.73	5.6	–
Ba	–	60	–	3.5	18	–
Be	194	3.3	–	1.85	12	–
Bi	9	117	–3.3	9.80	13.4	9
Cd	103	7.3	4.7	8.64	31	20
Ca	125	3.7	–	1.54	22	13
C	16.3	–	–	2.30	7.9	–
Ce	11.9	85.4	–	6.75	8	–
Cr	91.3	13.2	–	7.10	6.5	350
Co	96	6.3	–	8.90	12.5	125
Cu	397	1.69	4.1	8.96	17	48
Ga	41	–	–	5.91	18.3	–
Au	316	2.2	5.2	19.3	14.1	18.5
In	80	8.8	–	7.3	24.8	1
Ir	147	5.1	–	22.4	6.8	172
Fe	78	10.1	5.5	7.87	12.1	66
Pb	35	20.6	3.4	11.68	29	5.5
Li	76	9.3	1.5	0.53	56	–
Mg	156	4.2	4.2	1.74	26	25
Mn	7.8	160	–	7.4	23	–
Hg	8.7	96	3.75	13.55	61	–
Mo	137	5.7	–	10.2	5.1	147
Ni	89	6.9	–	8.9	13.3	80
Nb	54	16	–	8.6	7.2	–
Os	87	8.8	–	22.5	4.6	–
Pd	75	10.8	–	12.0	11.0	50
P	–	–	–	1.83	6.2	–
Pt	73	10.6	–	21.45	9.0	52
K	104	6.8	2.8	0.86	83	0.04
Rh	148	4.7	–	12.4	8.5	156
Si	139	10^3–10^6	–	2.34	7.6	–
Ag	425	1.6	4.5	10.5	19.1	25
Na	128	4.7	2.5	0.97	71	0.1
Sr	–	23	–	2.6	100	–
S	272	–	–	2.07	70	–
Ta	58	13.5	–	16.6	6.5	40
Te	3.8	1.6×10^5	–	6.24	–	–
Tl	45.5	16.6	–	11.85	30	–
Sn	73.2	12.6	2.8	7.3	23.5	–
Ti	21.6	54	–	4.5	8.9	–
W	174	5.4	–	19.3	4.5	–
U	28	??	–	19.0	–	–
V	31.6	19.6	–	6.1	8.3	–
Zn	120	6.0	6.5	7.14	31	35
Zr	22.6	44	–	6.49	5.9	–

Densities of casting alloys

Alloy	BS1490	g/ml	Alloy	BS1400	g/ml
Aluminium alloys			*Copper alloys*		
Pure Al		2.70	HC copper	HCC1	8.9
Al–Si5Cu3	LM4	2.75	Brass CuZn38Al	DCB1	8.5
Al–Si7Mg	LM25	2.68	CuZn33Pb2Si	HTB1	8.5
Al–Si8Cu3Fe	LM24	2.79	CuZn33Pb2	SCB3	8.5
AlSi12	LM6	2.65	Phosphor bronze		
			CuSn11P	PB1	8.8
Cast steels			CuSn12	PB2	8.7
Low carbon <0.20		7.86	Lead bronze		
Med. carbon 0.40		7.86	CuSn5Pb20	LB5	9.3
High carbon >0.40		7.84	Al bronze		
			CuAl10Fe2	AB1	7.5
Low alloy		7.86	Gunmetal		
Med. alloy		7.78	CuSnPb5Zn5	LG2	8.8
Med./high alloy		7.67	Copper nickel		
			CuNi30Cr2FeMnSi	CN1	8.8
Stainless					
13Cr		7.61	*Cast irons*		
18Cr8Ni		7.75	Grey iron 150 MPa		6.8–7.1
			200		7.0–7.2
Other alloys			250		7.2–7.4
Zinc base			300		7.3–7.4
ZnAl4Cu1		6.70	Whiteheart malleable		7.45
			Blackheart malleable		7.27
Lead base			White iron		7.70
PbSb6		10.88	Ductile iron (s.g.)		7.2–7.3
Tin base (Babbit)		7.34	Ni-hard		7.6–7.7
Inconel Ni76Cr18		8.50	High silicon (15%)		6.8

Approximate bulk densities of common materials

Material	kg/m³	lb/ft³	Material	kg/m³	lb/ft³
Aluminium, cast	2560	160	Lead	11370	710
wrought	2675	167	Limestone	2530–2700	158–168
Aluminium bronze	7610	475			
Ashes	590	37	Magnesite	2530	158
			Mercury	13560	847
Brass, rolled	8390	524	Monel	8870	554
swarf	2500	157			
Babbit metal	7270	454	Nickel, cast	8270	516
Brick, common	1360–1890	85–118	Nickel silver	8270	516
fireclay	1840	115			
Bronze	8550	534	Phosphor bronze	8580	536
			Pig iron, mean	4800	300
Cast iron, solid	7210	450	Pig iron and scrap		
turnings	2240	140	(cupola charge)	5400	336
Cement, loose	1360	85			
Chalk	2240	140	Sand, moulding	1200–1440	75–90
Charcoal, lump	290	18	silica	1360–1440	85–90
Clay	1900–2200	120–135	Silver, cast	10500	656
Coal	960–1280	60–80	Steel	7850	490
Coal dust	850	53			
Coke	450	28	Tin	7260	453
Concrete	2240	140			
Copper, cast	8780	548	Water, ice	940	58.7
Cupola slag	2400	150	liquid 0°C	1000	62.4
			100°C	955	59.6
Dolomite	2680	167	Wood, balsa	100–130	7–8
			oak	830	52
Fire clay	1440	90	pine	480	30
French chalk	2600	162	teak	640	40
			Wrought iron	7700	480
Glass	2230	139			
Gold, pure	19200	1200	Zinc, cast	6860	428
22 carat	17500	1090	rolled	7180	448
Graphite, powder	480	30			
solid	2200	138			

Patternmakers' contraction allowances

Castings are always smaller in dimensions than the pattern from which they are made, because as the metal cools from its solidification temperature to room temperature, thermal contraction occurs. Patternmakers allow for this contraction by making patterns larger in dimensions than the required castings by an amount known as the "contraction allowance". Originally this was done by making use of specially engraved rules, known as "contraction rules", the dimensions of which incorporated a contraction allowance such as 1 in 75 for aluminium alloys, or 1 in 96 for iron castings. Nowadays, most patterns and coreboxes are made using computer-controlled machine tools and it is more convenient to express the contraction as a percentage allowance.

Predicting casting contraction can never be precise, since many factors are involved in determining the exact amount of contraction that occurs. For example, when iron castings are made in greensand moulds, the mould walls may move under the pressure of the liquid metal, causing expansion of the mould cavity, thus compensating for some of the metal contraction. Cored castings may not contract as much as expected, because the presence of a strong core may restrict movement of the casting as it is cooling. Some core binders expand with the heat of the cast metal causing the casting to be larger than otherwise expected. For these reasons, and others, it is only possible to predict contractions approximately, but if a patternmaker works with a particular foundry for a long period, he will gain experience with the foundry's design of castings and with the casting methods used in the foundry. Based on such experience, more precise contraction allowances can be built into the patterns.

The usually accepted contraction allowances for different alloys are given in the following table.

Alloy		Contraction allowance (%)
Aluminium alloys		
Al–Si5Cu3	LM4	
Al–Si7Mg	LM25	1.3
Al–Si8Cu3Fe	LM24	
Al–Si12	LM6	
Beryllium copper		1.6
Bismuth		1.3
Brass		1.56
Bronze, aluminium		2.32
manganese		0.83–1.56
phosphor		1.0–1.6
silicon		1.3–1.6
Cast iron, grey		0.9–1.04
white		2.0
ductile (s.g.)		0.6–0.8
malleable		1.0–1.4
Copper		1.6
Gunmetal		1.0–1.6
Lead		2.6
Magnesium alloys		1.30–1.43
Monel		2.0
Nickel alloys		2.0
Steel, carbon		1.6–2.0
chromium		2.0
manganese		1.6–2.6
Tin		2.0
White metal		0.6
Zinc alloys		1.18

Volume shrinkage of principal casting alloys

Most alloys shrink in volume when they solidify, the shrinkage can cause voids in castings unless steps are taken to "feed" the shrinkage by the use of feeders.

Casting alloy	Volume shrinkage (%)
Carbon steel	6.0
Alloyed steel	9.0
High alloy steel	10.0
Malleable iron	5.0
Al	8.0
Al–Cu4Ni2Mg	5.3
Al–Si12	3.5
Al–Si5Cu2Mg	4.2
Al–Si9Mg	3.4
Al–Si5Cu1	4.9
Al–Si5Cu2	5.2
Al–Cu4	8.8
Al–Si10	5.0
Al–Si7NiMg	4.5
Al–Mg5Si	6.7
Al–Si7Cu2Mg	6.5
Al–Cu5	6.0
Al–Mg1Si	4.7
Al–Zn5Mg	4.7
Cu (pure)	4.0
Brass	6.5
Bronze	7.5
Al bronze	4.0
Sn bronze	4.5

Comparison of sieve sizes

Sieves used for sand grading are of 200 mm diameter and are now usually metric sizes, designated by their aperture size in micrometres (μm). The table lists sieve sizes in the British Standard Metric series (BS410:1976) together with other sieve types.

Sieve aperture, micrometres and sieve numbers

ISO/R.565 series (BS410:1976) (μm)	BSS		ASTM	
	No.	μm	No.	μm
(1000)	16	1003	18	1000
710	22	699	22	710
500	30	500	30	500
355	44	353	45	350
250	60	251	60	250
(212)	72	211	70	210
180				
(150)	100	152	100	149
125			120	125
90	150	104	150	105
63	200	76	200	74
(45)	300	53	325	44

Notes: The 1000 and 45 sieves are optional.
The 212 and 150 sieves are also optional, but may be included to give better separation between the 250 and 125 sieves.

Calculation of average grain size

The adoption of the ISO metric sieves means that the old AFS grain fineness number can no longer be calculated. Instead, the average grain size, expressed as micrometres (μm) is now used. This is determined as follows:

1 Weigh a 100 g sample of dry sand.
2 Place the sample into the top sieve of a nest of ISO sieves on a vibrator. Vibrate for 15 minutes.
3 Remove the sieves and, beginning with the top sieve, weigh the quantity of sand remaining on each sieve.
4 Calculate the percentage of the sample weight retained on each sieve, and arrange in a column as shown in the example.
5 Multiply the percentage retained by the appropriate multiplier and add the products.
6 Divide by the total of the percentages retained to give the average grain size.

Example

ISO aperture (μm)	Percentage retained	Multiplier	Product
≥710	trace	1180	0
500	0.3	600	180
355	1.9	425	808
250	17.2	300	5160
212	25.3	212	5364
180	16.7	212	3540
150	19.2	150	2880
125	10.6	150	1590
90	6.5	106	689
63	1.4	75	105
≤63	0.5	38	19
Total	99.6	–	20 335

Average grain size = 20 335/99.6
= 204 μm

Calculation of AFS grain fineness number

Using either the old BS sieves or AFS sieves, follow, steps 1–4 above.
5 Arrange the results as shown in the example below.
6 Multiply each percentage weight by the preceding sieve mesh number.
7 Divide by the total of the percentages to give the AFS grain fineness
  number.

Example

BS sieve number	% sand retained on sieve	Multiplied by previous sieve no.	Product
10	nil	–	–
16	nil	–	–
22	0.2	16	3.2
30	0.8	22	17.6
44	6.7	30	201.0
60	22.6	44	1104.4
100	48.3	60	2898.0
150	15.6	100	1560.0
200	1.8	150	270.0
pan	4.0	200	800.0
Total	100.0	–	6854.2

AFS grain fineness number = 6854.2/100
                           = 68.5 or 68 AFS

Foundry sands usually fall into the range 150–400 μm, with 220–250 μm
being the most commonly used. Direct conversion between average grain
size and AFS grain fineness number is not possible, but an approximate
relation is shown below:

AFS grain fineness no.	35	40	45	50	55	60	65	70	80	90
Average grain size (μm)	390	340	300	280	240	220	210	195	170	150

While average grain size and AFS grain fineness number are useful
parameters, choice of sand should be based on particle size distribution.

Recommended standard colours for patterns

Part of pattern		*Colour*
As-cast surfaces which are to be left unmachined		Red or orange
Surfaces which are to be machined		Yellow
Core prints for unmachined openings and end prints	Periphery	Black
	Ends	Black
Core prints for machined openings	Periphery	Yellow stripes on black
	Ends	Black
Pattern joint (split patterns)	Cored section	Black
	Metal section	Clear varnish
Touch core	Cored shape	Black
	Legend	"Touch"
Seats of and for loose pieces and loose core prints		Green
Stop offs		Diagonal black stripes with clear varnish
Chilled surfaces	Outlined in	Black
	Legend	"Chill"

Dust control in foundries

Air extraction is used in foundries to remove silica dust from areas occupied by operators. The following table indicates the approximate air velocities needed to entrain sand particles.

Terminal velocities of spherical particles of density 2.5 g/cm^3 (approx.)

BS sieve size	Particle dia. (μm)	Terminal velocity m/sec	ft/sec	ft/min
16	1003	7.0	23	1400
30	500	4.0	13	780
44	353	2.7	9	540
60	251	1.8	6	360
100	152	1.1	3.5	210
150	104	0.5	1.7	100
200	76	0.4	1.3	80

For the comfort and safety of operators, air flows of around 0.5 m/sec are needed to carry away silica dust. If air flow rate is too high, around the shake-out for example, there is a danger that the grading of the returned sand will be altered.

Buoyancy forces on cores

When liquid metal fills a mould containing sand cores, the cores tend to float and must be held in position by the core prints or by chaplets. The following table lists the buoyancy forces experienced by silica sand cores in various liquid metals, expressed as a proportion of the weight of the core:

Liquid metal	Ratio of buoyant force to core weight
Aluminium	0.66
Brass	4.25
Copper	4.50
Cast iron	3.50
Steel	3.90

Core print support

Moulding sand (green sand) in a core print will support about $150\,kN/m^2$ (21 psi). So the core print can support the following load:

Support (kN) = Core print area $(m^2) \times 150$

1 kN = 100 kgf (approx.)

Support (kgf) = Core print area $(m^2) \times 15\,000$

Example: A core weighing 50 kg has a core print area of 10×10 cm (the area of the upper, support surface), i.e. $0.1 \times 0.1 = 0.01\,m^2$. The print support is $150 \times 0.01 = 1.5\,kN = 150\,kgf$

If the mould is cast in iron, the buoyancy force is $50 \times 3.5 = 175\,kgf$ so chaplets may be necessary to support the core unless the print area can be increased.

Opening forces on moulds

Unless a mould is adequately clamped or weighted, the force exerted by the molten metal will open the boxes and cause run-outs. If there are insufficient box bars in a cope box, this same force can cause other problems like distortion and sand lift. It is important therefore to be able to calculate the opening force so that correct weighting or clamping systems can be used.

The major force lifting the cope of the mould is due to the metallostatic pressure of the molten metal. This pressure is due to the height, or head, of metal in the sprue above the top of the mould (H in Fig. 1.1). Additional

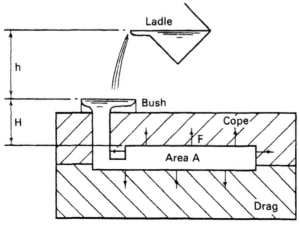

Figure 1.1   *Opening forces of moulds.*

forces exist from the momentum of the metal as it fills the mould and from forces transmitted to the cope via the core prints as the cores in cored castings try to float.

The momentum force is difficult to calculate but can be taken into account by adding a 50% safety factor to the metallostatic force.

The opening metallostatic force is calculated from the total upward-facing area of the cope mould in contact with the metal (this includes the area of all the mould cavities in the box). The force is:

$$F(kgf) = \frac{A \times H \times d \times 1.5}{1000}$$

A is the upward facing area in cm^2
H (cm) is the height of the top of the sprue above the average height of the upward facing surface
d is the density of the molten metal (g/cm^3)
1.5 is the "safety factor"

For ferrous metals, d is about 7.5, so:

$$F(kgf) = \frac{11 \times A \times H}{1000}$$

For aluminium alloys, d is about 2.7, so:

$$F(kgf) = \frac{4 \times A \times H}{1000}$$

Forces on cores

The core tends to float in the liquid metal and exerts a further upward force (see page 18)

In the case of ferrous castings, this force is

$$3.5 \times W \text{ (kgf)}$$

where W is the weight of the cores in the mould (in kg).
In aluminium, the floating force can be neglected.
The total resultant force on the cope is (for ferrous metals)

$$(11 \times A \times H)/1000 + 3.5 W \text{ kgf}$$

Example: Consider a furane resin mould for a large steel valve body casting having an upward facing area of 2500 cm² and a sprue height (*H*) of 30 cm with a core weighing 40 kg. The opening force is

$$11 \times 2500 \times 30/1000 + 3.5 \times 40 = 825 + 140$$

$$= 965 \, kgf$$

It is easy to see why such large weights are needed to support moulds in jobbing foundries.

Dimensional tolerances and consistency achieved in castings

Errors in dimensions of castings are of two kinds:

Accuracy: the variation of the mean dimension of the casting from the design dimension given on the drawing

Consistency: statistical errors, comprising the dimensional variability round the mean dimension

Dimensional accuracy

The major causes of deviations of the mean dimension from the target value are contraction uncertainty and errors in pattern dimensions. It is usually possible to improve accuracy considerably by alternations to pattern equipment after the first sample castings have been made.

Dimensional consistency

Changes in process variables during casting give rise to a statistical spread of measurements about a mean value. If the mean can be made to coincide with the nominal dimension by pattern modification, the characteristics of this statistical distribution determine the tolerances feasible during a production run.

The consistency of casting dimensions is dependent on the casting process used and the degree of process control achieved in the foundry. Fig. 1.2 illustrates the average tolerance exhibited by various casting processes. The tolerance is expressed as 2.5σ (2.5 standard deviations), meaning that only 1 casting in 80 can be expected to have dimensions outside the tolerance.

There is an International Standard, ISO 8062–1984(E) *Castings – System of dimensional tolerances,* which is applicable to the dimensions of cast metals and their alloys produced by sand moulding, gravity diecasting, low

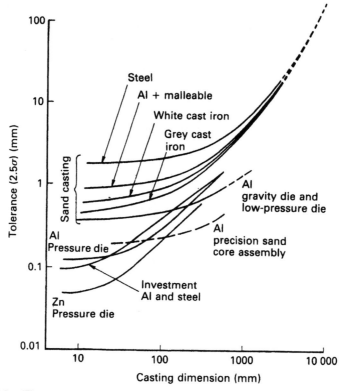

Figure 1.2   *The average tolerance (taken as 2.5σ) exhibited by various casting processes. (From Campbell, J. (1991) Castings, Butterworth-Heinemann, reproduced by permission of the publishers.)*

pressure diecasting, high pressure dicasting and investment casting. The Standard defines 16 tolerance grades, designated CT1 to CT16, listing the total casting tolerance for each grade on raw casting dimensions from 10 to 10 000 mm. The Standard also indicates the tolerance grades which can be expected for both long and short series production castings made by various processes from investment casting to hand-moulded sand cast.

Reference should be made to ISO 8062 or the equivalent British Standard BS6615:1985 for details.

Chapter 2
Aluminium casting alloys

Introduction

Aluminium casting is dominated by the automotive industry. Roughly two thirds of all aluminium castings are automotive where the use of aluminum castings continues to grow at the expense of iron castings. Although aluminium castings are significantly more expensive than ferrous castings, there is a continuing market requirement to reduce vehicle weight and to increase fuel efficiency. It is this requirement which drives the replacement of ferrous parts by aluminium.

Aluminium castings are widely used in cars for engine blocks, heads, pistons, rocker covers, inlet manifolds, differential casings, steering boxes, brackets, wheels etc. The potential for further use of aluminium in automotive applications is considerable. European cars in 1992 had 50–60 kg Al castings and this is expected to double by year 2000.

When aluminium alloys are cast, there are many potential sources of defects which can harm the quality of the cast part. All aluminium alloys are subject to:

Shrinkage defects	Al alloys shrink by 3.5–6.0% during solidification (depending on alloy type)
Gas porosity	Molten aluminium readily picks up hydrogen which is expelled during solidification giving rise to porosity
Oxide inclusions	Molten Al exposed to air immediately oxidises forming a skin of oxide which may be entrained into the casting

Because of these potential problems aluminium castings, like all castings, suffer from variable mechanical properties which can be described by a distribution curve. The mechanical properties used by the designer of the casting must take the distribution curve into account. If, for example, the process mean tensile strength for a cast alloy is 200 MPa, the designer must use a lower figure, say 150 MPa, as the strength of the alloy to take into account the variability of properties. If the spread of the distribution curve can be reduced, then a higher design strength, say 170 MPa can be used, even though the process mean for the alloy and the casting process stays the same.

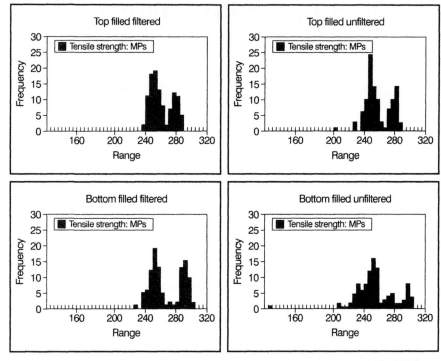

Figure 2.1   *Histograms showing the distribution of tensile strength of Al–Si7Mg alloy test bars cast in various ways. (From Foseco Foundry Practice, **226**, July 1995.)*

The strength of castings does not follow the normal bell-shaped distribution curve. Figure 2.1 shows the range of tensile strengths found in 12.5 mm diameter test bars cast in an Al–Si7 Mg alloy into resin bonded sand moulds using various pouring methods: top or bottom filled, filtered or unfiltered. In all cases the process mean tensile strength is about 260 MPa, but the distribution is different.

> The unfiltered castings show a few but very significant low strength test pieces, known as outliers
> For each filling category the plots show two distinct bands of tensile strength.

A design strength below 200 MPa would have to be used for unfiltered castings because of the occasional outliers.

Examination of the fracture surface of the low strength outliers showed massive oxide fragments indicating that inclusions in the unfiltered castings were responsible for the low tensile strength. Filtration of the metal eliminates the inclusions allowing the design strength to be increased to around 230 MPa.

The double band appearance of the histograms is interpreted as indicating that more than one defect type is acting to control the behaviour at fracture.

It is by reducing the variability of properties of castings that the greatest progress has been made in recent years. This has allowed designers to have greater confidence in castings so that thinner sections and lower weight components can be used. The stages in the aluminium casting process where the greatest improvements have been made are:

Efficient degassing
Grain refinement
Modification of structure
Metal filtration
Non-turbulent filling of moulds

Chill casting (into metal moulds) has inherently a greater possibility of producing higher quality than sand casting because the higher rate of solidification reduces pore size and refines grain size. The highest quality components are produced using filtered metal, non-turbulently introduced into metal moulds and solidified under high external pressure to minimise or totally avoid porosity. While it is not always possible to use high external pressure during solidification (castings using sand cores will suffer from metal penetration), the understanding of the origins of defects in aluminium castings and their reduction by attention to degassing, metal treatment and filtration has greatly improved the general quality of castings in recent years. There is little doubt that improvements will continue to be made in the future.

Casting alloys

There is a large and confusing range of casting alloys in use worldwide, defined by the National Specifications of the major industrial countries. Unfortunately there is little correspondence between the Standard Alloys used in different countries.

A European Standard for Aluminium Casting Alloys, EN 1706, was approved in August 1997 and the English language version BS EN 1706:1998 was published in March 1998. Along with the following standards, it partially supersedes BS 1490:1988 which will be withdrawn when EN 1559–4 is published.

BS EN 1559–1:1997	Founding. Technical conditions of delivery. General
BS EN 1676:1997	Aluminium and aluminium alloys. Alloyed ingots for remelting. Specifications
PrEN 1559–4	Founding. Technical conditions of delivery. Additional requirements for aluminium castings

BS EN 1706:1998 specifies the chemical compositions of 37 alloys. For each alloy, mechanical properties are specified only for the commonly used methods of casting and for commonly used tempers. Refer to BS EN 1706:1988 for full details.

Tables 2.1a, b, c and d list the alloy designation of alloys commonly used for (a) sand casting, (b) chill casting, (c) pressure diecasting and (d) investment casting.

Table 2.2 lists the chemical composition of some commonly used BS EN 1706 alloys and their equivalent BS 1490 alloys.

Table 2.3 lists mechanical properties specified for the alloys in Table 2.2.

Many foundries are still unfamiliar with the European Standard and have not yet converted from the National Standards.

Table 2.1a   BS EN 1706:1988 alloys commonly used for sand casting

Alloy group	Alloy designation	
	Numerical	*Chemical symbols*
Al Cu	EN AC-21000	EN AC-Al Cu4MgTi
	EN AC-21100	EN AC-Al Cu4Ti
Al SiMgTi	EN AC-41000	EN AC-Al Si2MgTi
Al Si7Mg	EN AC-42000	EN AC-Al Si7Mg
	EN AC-42100	EN AC-Al Si7Mg0,3
	EN AC-42200	EN AC-Al Si7Mg0,6
Al Si10M	EN AC-43000	EN AC-Al Si10Mg(a)
	EN AC-43100	EN AC-Al Si10Mg(b)
	EN AC-43200	EN AC-Al Si10Mg(Cu)
	EN AC-43300	EN AC-Al Si9Mg
Al Si	EN AC-44000	EN AC-Al Si11
	EN AC-44100	EN AC-Al Si12(b)
	EN AC-44200	EN AC-Al Si12(a)
Al Si5Cu	EN AC-45000	EN AC-Al Si6Cu4
	EN AC-45200	EN AC-Al Si5Cu3Mn
	EN AC-45300	EN AC-Al Si5Cu1Mg
Al Si9Cu	EN AC-46200	EN AC-Al Si8Cu3
	EN AC-46400	EN AC-Al Si9Cu1Mg
	EN AC-46600	EN AC-Al Si7Cu2
Al Si(Cu)	EN AC-47000	EN AC-Al Si12(Cu)
Al Mg	EN AC-51000	EN AC-Al Mg3(b)
	EN AC-51100	EN AC-Al Mg3(a)
	EN AC-51300	EN AC-Al Mg5
	EN AC-51400	EN AC-Al Mg5(Si)
Al ZnMg	EN AC-71000	EN AC-Al Zn5Mg

Table 2.1b   BS EN 1706:1988 alloys commonly used for chill casting

Alloy group	Alloy designation	
	Numerical	*Chemical symbols*
AlCu	EN AC-21000	EN AC-Al Cu4MgTi
	EN AC-21100	EN AC-Al Cu4Ti
AlSiMgTi	EN AC-41000	EN AC-Al Si2MgTi
AlSi7Mg	EN AC-42000	EN AC-Al Si7Mg
AlSi10Mg	EN AC-42100	EN AC-Al Si7Mg0,3
	EN AC-42200	EN AC-Al Si7Mg0,6
	EN AC-43000	EN AC-Al Si10Mg(a)
	EN AC-43100	EN AC-Al Si10Mg(b)
	EN AC-43200	EN AC-Al Si10Mg(Cu)
	EN AC-43300	EN AC-Al Si9Mg
AlSi	EN AC-44000	EN AC-Al Si11
	EN AC-44100	EN AC-Al Si12(b)
	EN AC-44200	EN AC-Al Si12(a)
AlSi5Cu	EN AC-45000	EN AC-Al Si6Cu4
	EN AC-45100	EN AC-Al Si5Cu3Mg
	EN AC-45200	EN AC-Al Si5Cu3Mn
	EN AC-45300	EN AC-Al Si5Cu1Mg
	EN AC-45400	EN AC-Al Si5Cu3
AlSi9Cu	EN AC-46200	EN AC-Al Si8Cu3
	EN AC-46300	EN AC-Al Si7Cu3Mg
	EN AC-46400	EN AC-Al Si9Cu1Mg
	EN AC-46600	EN AC-Al Si7Cu2
AlSi(Cu)	EN AC-47000	EN AC-Al Si12(Cu)
AlSiCuNiMg	EN AC-48000	EN AC-Al Si12CuNiMg
AlMg	EN AC-51000	EN AC-Al Mg3(b)
	EN AC-51100	EN AC-Al Mg3(a)
	EN AC-51300	EN AC-Al Mg5
	EN AC-51400	EN AC-Al Mg5(Si)
AlZnMg	EN AC-71000	EN AC-Al Zn5Mg

Table 2.4 lists the BS 1490:1988 "LM alloys" and their approximate equivalents in European, National and International Standards.

Table 2.5 shows the chemical composition of the LM alloys.

Table 2.6 gives the specified minimum mechanical properties. Refer to BS 1490:1988 for details of test bar dimensions, details of testing and heat treatment methods.

Table 2.7 lists some British Standard alloys used for aerospace applications and their equivalents used in other countries.

Table 2.1c   BS EN 1706:1988 alloys commonly used for pressure diecasting

Alloy group	Alloy designation	
	Numerical	Chemical symbols
AlSi10Mg	EN AC-43400	EN AC-Al Si10Mg(Fe)
AlSi	EN AC-44300	EN AC-Al Si12(Fe)
	EN AC-44400	EN AC-Al Si9
AlSi9Cu	EN AC-46000	EN AC-Al Si9Cu3(Fe)
	EN AC-46100	EN AC-Al Si11Cu2(Fe)
	EN AC-46200	EN AC-Al Si8Cu3
	EN AC-46500	EN AC-Al Si9Cu3(Fe)(Zn)
AlSi(Cu)	EN AC-47100	EN AC-Al Si12Cu1(Fe)
AlMg	EN AC-51200	EN AC-Al Mg9

Table 2.1d   BS EN 1706:1988 alloys commonly used for investment casting

Alloy group	Alloy designation	
	Numerical	Chemical symbols
AlCu	EN AC-21000	EN AC-Al Cu4MgTi
AlSi7Mg	EN AC-42000	EN AC-Al Si7Mg
	EN AC-42100	EN AC-Al Si7Mg0,3
	EN AC-42200	EN AC-Al Si7Mg0,6
AlSi	EN AC-44100	EN AC-Al Si12(b)
AlSi5Cu	EN AC-45200	EN AC-Al Si5Cu3Mn
AlMg	EN AC-51300	EN AC-Al Mg5

Table 2.8 summarises the casting characteristics of BS 1490 alloys.
Table 2.9 lists their common uses.
Table 2.10 gives the colour codes used for BS 1490 ingots.

These tables are based on BS EN 1706:1998 and on tables in British and European Aluminium Casting Alloys, their Properties and characteristics. ALARS Ltd. (1996) Thanks are due to the British Standards Institution and the Association of Light Alloy Refiners, Birmingham, for permission to use them.

Table 2.2 Commonly used BS EN 1706:1998 alloys

Alloy designation

Numerical	Chemical symbols	Si	Fe	Cu	Mn	Mg	Cr	Ni	Zn	Pb	Sn	Ti	Others Each	Others Total	Equivalent BS 1490 alloy	Casting process
EN AC-42000	EN AC-Al Si7Mg	6.5–7.5	0.55	0.2	0.35	0.2–0.65	–	0.15	0.15	0.15	0.05	0.05–0.25	0.05	0.15	LM25	sand/chill
EN AC-44100	EN AC-Al Si12(b)	10.5–13.5	0.65	0.15	0.55	0.1	–	0.1	0.15	0.1	–	0.2	0.05	0.15	LM6	sand/chill
EN AC-45200	EN AC-Al Si5Cu3Mn	4.5–6.0	0.8	2.5–4.0	0.20–0.55	0.4	–	0.3	0.55	0.2	0.1	0.2	0.05	0.25	LM4	sand/chill
EN AC-46600	EN AC-Al Si7Cu2	6.0–8.0	0.8	1.5–2.5	0.15–0.65	0.35	–	0.35	1	0.25	0.15	0.25	0.05	0.15	LM27	sand/chill
EN AC-46500	EN AC-Al Si9Cu3(Fe)(Zn)	8.0–11.0	1.3	2.0–4.0	0.55	0.05–0.55	0.15	0.55	3	0.35	0.25	0.25	0.05	0.25	LM24	pressure d/c

Single figure are maximum.

Table 2.3 Mechanical properties of commonly used BS EN 1706:1998 alloys

Alloy designation Numerical	Chemical symbols	Casting method	Temper	Tensile strength (Mpa)	Proof stress (Mpa)	Elong. (%)	Brinell HBS	Equivalent BS 1490 alloy
EN AC-42000	AlSi7Mg	sand	F	140	80	2	50	LM25
		sand	T6	220	180	1	75	
		chill	F	170	90	2.5	55	
		chill	T6	260	220	1	90	
		chill	T64	240	200	2	80	
EN AC-44100	AlSi12(b)	sand	F	150	70	4	50	LM6
		chill	F	170	80	5	55	
EN AC-45200	AlSi5Cu3Mn	sand	F	140	70	1	60	LM4
		sand	T6	230	200	<1	90	
		chill	F	160	80	1	70	
		chill	T6	280	230	<1	90	
EN AC-46600	AlSi7Cu2	sand	F	150	90	1	60	LM27
		chill	F	170	100	1	75	
EN AC-46500*	AlSi9Cu3(Fe)(Zn)	pressure d/c	F	240	140	<1	80	LM24

*Mechanical properties for guidance only.

Table 2.4 BS 1490:1988 alloys and approximate equivalents

UK	ISO	EN AC-	France	Germany	Italy UNI	USA AA/ASTM	USA SAE	Japan	End uses
LM0	Al 99.5	–	A5	–	3950	150	–	–	Electrical, food, chemical plant
LM2	Al-Si10Cu2Fe	46 100	A-S9U3-Y4	–	5076	384	383	ADC 12	Pressure diecasting
LM4	Al-Si5Cu3	45 200	A-S5U3	G-AlSi6Cu4 (225)	3052	319	326	AC 2A	Sand, gravity diecast manifolds, gear boxes etc.
LM5	Al-Mg5Si1 / Al-Mg6	51 300	AG6	G-AlMg5 (244)	3058	514	320	AC 7A	Sand, gravity; corrosion resistant, for marine use
LM6	Al-Si12 / Al-Si12Fe	44 100	AS13	G-AlSi12 (230)	4514	A413	–	AC 3A	Food plant, chemical plant; Sand, gravity; thin sections, manifolds etc.
LM9	Al-Si10Mg	43 100	A-S10G	G-AlSi10Mg (233)	3049	A360	309	AC 4A	Low pressure etc.; motor housings, cover plates etc. High strength when heat treated
LM12	Al-Cu10Si2Mg	–	A-U10G	–	3041	222	34	–	Gravity, sand cast; machines well, hydraulic equipment
LM13	Al-Si12Cu / Al-Si12CuFe	48 000	A-S12UN	–	3050	336	321	AC 8A	Sand, chill; used for pistons
LM16	Al-Si5Cu1Mg	45 300	A-S4UG	–	3600	355	322	AC 4D	Sand, chill; cylinder heads valve bodies, good pressure tightness
LM20	Al-Si12Cu / Al-Si12CuFe	47 000	A-S12-Y4	G-AlSi12(Cu) (231)	5079	A413	305	–	Pressure diecasting corrosion resistant, marine castings, water pumps, meter cases
LM21	Al-Si6Cu4	45 000	A-S5U2	G-AlSi6Cu4 (225)	7369/4	308	326	AC 2A	Sand, gravity; similar to LM4, crankcases, gear boxes etc.
LM22	Al-Si5Cu3	45 400	A-S5U	G-AlSi6Cu4 (225)	3052	319	326	AC 2A	Chill casting; solution treated, good shock resistance, automotive heavy duty parts
LM24	Al-Si8Cu3Fe	46 500	A-S9U3A-Y4	G-AlSi8Cu3 (226)	5075 / 3601	A380	306	AC 4B / ADC10	Pressure diecasting; engineering diecastings
LM25	Al-Si7Mg	42 000	A-S7G	G-AlSi7Mg	3599	A356	323	AC 4C	Sand, chill; general purpose high strength alloy with good castability; wheels, cylinder blocks, heads
LM26	Al-Si9Cu3Mg	–	A-S7U3G	–	3050	332	332	–	Chill; used for pistons
LM27	Al-Si7Cu2Mn0.5	46 600	–	–	7369	–	–	AC 2B	Sand, chill; versatile alloy, good castability, general engineering parts
LM28	Al-Si19CuMgNi	–	–	–	6251	–	–	–	Chill; high performance pistons
LM29	Al-Si23CuMgNi	–	–	–	6251	–	–	–	Chill; high performance pistons
LM30	Al-Si17Cu4Mg	–	–	–	–	390	–	–	Pressure diecast; unlined cylinder blocks
LM31	Al-Zn5Mg	71 000	A-Z5G	–	3602	712	310	–	Sand; large castings, good shock resistance, good strength at elevated temp.

Table 2.5 Chemical composition (weight per cent) of BS 1490:1988 alloys

Alloy	Cu	Mg	Si	Fe	Mn	Ni	Zn	Pb	Sn	Ti	Additional elements	Others Each	Total
LM0	0.03	0.03	0.30	0.40	0.03	0.03	0.07	0.03	0.03	—	Al 99.50 min	—	—
LM2	0.7–2.5	0.30	9.0–11.5	1.0	0.5	0.5	2.0	0.3	0.2	0.2	—	—	0.50
LM4	2.0–4.0	0.20	4.0–6.0	0.8	0.2–0.6	0.3	0.5	0.1	0.1	0.2	—	0.05	0.15
LM5	0.1	3.0–6.0	0.3	0.6	0.3–0.7	0.1	0.1	0.05	0.05	0.2	—	0.05	0.15
LM6	0.1	0.10	10.0–13.0	0.6	0.5	0.1	0.1	0.1	0.05	0.2	—	0.05	0.15
LM9	0.20	0.2–0.6	10.0–13.0	0.6	0.3–0.7	0.1	0.1	0.1	0.05	0.2	—	0.05	0.15
LM12	9.0–11.0	0.2–0.4	2.5	1.0	0.6	0.5	0.8	0.1	0.1	0.2	—	0.05	0.15
LM13	0.7–1.5	0.8–1.5	10.0–13.0	1.0	0.5	1.5	0.5	0.1	0.1	0.2	—	0.05	0.15
LM16	1.0–1.5	0.4–0.6	4.5–5.5	0.6	0.5	0.25	0.1	0.1	0.05	0.2¹	—	0.05	0.15
LM20	0.4	0.2	10.0–13.0	1.0	0.5	0.1	0.2	0.1	0.1	0.2	—	0.05	0.20
LM21	3.0–5.0	0.1–0.3	5.0–7.0	1.0	0.2–0.6	0.3	2.0	0.2	0.1	0.2	—	0.05	0.15
LM22	2.8–3.8	0.05	4.0–6.0	0.6	0.2–0.6	0.15	0.15	0.1	0.05	0.2	—	0.05	0.15
LM24	3.0–4.0	0.30	7.5–9.5	1.3	0.5	0.5	3.0	0.3	0.2	0.2	—	—	0.50
LM25	0.20	0.20–0.6	6.5–7.5	0.5	0.3	0.1	0.1	0.1	0.05	0.2¹	—	0.05	0.15
LM26	2.0–4.0	0.5–1.5	8.5–10.5	1.2	0.5	1.0	1.0	0.2	0.1	0.2	—	0.05	0.15
LM27	1.5–2.5	0.35	6.0–8.0	0.8	0.2–0.6	0.3	1.0	0.2	0.1	0.2	—	0.05	0.15
LM28²	1.3–1.8	0.8–1.5	17–20	0.7	0.6	0.8–1.5	0.2	0.1	0.1	0.2	Cr 0.6 Co 0.5	0.10	0.30
LM29²	0.8–1.3	0.8–1.3	22–25	0.7	0.6	0.8–1.3	0.2	0.1	0.1	0.2	Cr 0.6 Co 0.5	0.10	0.30
LM30	4.0–5.0	0.4–0.7	16–18	1.1	0.3	0.1	0.2	0.1	0.1	0.2	—	0.10	0.30
LM31³	0.1	0.5–0.75	0.25	0.5	0.1	0.1	4.8–5.7	0.05	0.05	0.25¹	Cr 0.4–0.6	0.05	0.15

Notes Single figures in the table are maxima.
In cases where alloys are required in the modified condition, the level of any modifying element is not limited by the specified maximum value for "other elements".
1. 0.05% minimum if Ti alone is used for grain refining.
2. LM28 and LM29 castings are also subject to metallographic structure requirements.
3. LM31 castings in the M condition have to be naturally aged for 3 weeks before use or determination of mechanical properties. BS1490:1988 should be referred to for details.

Table 2.6 Typical mechanical properties of BS-1490:1998 alloys

Alloy condition	Sand cast				Chill/permanent mould cast					Alloy condition
	0.2% Proof stress (N/mm²)	Tensile stress (N/mm²)	Elong. (%)	Brinell hardness	0.2% Proof stress (N/mm²)	Tensile stress (N/mm²)	Elong. (%)	Brinell hardness	Strength at elevated temperatures	
LM0–M	30	80	30	25	30	80	40	25	—	LM0–M
LM2–M	—	—	—	—	90	180	2	80	G	LM2–M
LM4–M	100	150	2	70	100	200	3	80	G	LM4–M
LM4–TF	250	280	1	105	250	310	3	110	G	LM4–TF
LM5–M	90	170	5	65	90	230	10	65	F	LM5–M
LM6–M	70	170	8	55	80	200	13	60	P	LM6–M
LM9–M	—	—	—	—	—	200	3	—	G	LM9–M
LM9–TE	120	180	2	70	150	250	2.5	80	G	LM9–TE
LM9–TF	220	250	—	100	280	310	1	110	G	LM9–TF
LM12–M	—	—	—	—	150	180	—	95	G	LM12–M
LM13–TE	—	—	—	110	—	220	1	105	E	LM13–TE
LM13–TF	200	200	—	125	280	290	1	125	E	LM13–TF
LM13–TF7	140	150	1	77	200	210	1	75	E	LM13–TF7
LM16–TB	130	210	3	80	140	250	6	85	G	LM16–TB
LM16–TF	240	280	1	100	270	310	2	110	G	LM16–TF
LM20–M	—	—	—	—	80	220	7	60	P	LM20–M
LM21–M	130	180	1	85	130	200	2	90	G	LM21–M
LM22–TB	—	—	—	—	120	260	9	75	G	LM22–TB
LM24–M	—	—	—	—	110	200	2	85	G	LM24–M
LM25–M	90	140	2.5	60	90	180	5	60	F	LM25–M
LM25–TE	130	170	1.5	70	150	220	2	80	F	LM25–TE
LM25–TB7	100	170	3	65	100	230	8	65	F	LM25–TB7
LM25–TF	220	250	1	105	240	310	3	105	G	LM25–TF
LM26–TE	—	—	—	105	180	230	1	105	E	LM26–TE

Table 2.6 (Continued)

Alloy condition	Sand cast				Chill/permanent mould cast					Alloy condition
	0.2% Proof stress (N/mm²)	Tensile stress (N/mm²)	Elong. (%)	Brinell hardness	0.2% Proof stress (N/mm²)	Tensile stress (N/mm²)	Elong. (%)	Brinell hardness	Strength at elevated temperatures	
LM27–M	90	150	2	75	100	180	3	80	G	LM27–M
LM28–TE	—	—		110	170	190	0.5	120	G	LM28–TE
LM28–TF	120	130	0.5	120	170	200	0.5	120	G	LM28–TF
LM29–TE	120	130	0.5	120	170	210	0.3	120	G	LM29–TE
LM29–TF	120	130	0.5	120	170	210	0.3	120	G	LM29–TF
LM30–M	—	—	—	—	150	150	—	110	G	LM30–M
LM30–TS	—	—	—	—	160	160	—	110	G	LM30–TS
LM31–M	170	215	4	70	170	240	5	70	G	LM31–M
LM31–TE	170	215	4	70	170	240	5	70	G	LM31–TE

Notes:
1. Suffix letters indicate the condition of the casting: M = as-cast; TE = precipitation treated; TF = solution treated and precipitation treated; TF7 = fully heat treated and stabilised; TB7 = solution treated and stabilised; TS = stress relieved.
2. Strength at elevated temperatures: E = excellent; G = good; F = fair; P = poor. Diecasting alloys are restricted to moderately elevated temperatures.
3. The typical properties shown in this table are those of separately cast test bars and may not be obtained in all parts of the casting. Refer to BS1490:1988 for details.

Table 2.7 Aerospace alloy castings (British Standard L-series and Ministry of Defence Aerospace Alloys)

Alloy	Notes	Cu	Mg	Si	Fe	Mn	Ni	Zn	Pb	Sn	Ti	Others	
												Each	Total
BS2L99		0.10	0.20–0.45	6.5–7.5	0.20	0.10	0.10	0.10	0.05	0.05	0.20	—	—
BSL119	(a)	4.5–5.5	0.10	0.30	0.50	0.20–0.30	1.3–1.7	0.10	—	—	0.15–0.25	0.05	0.15
BSL154		3.8–4.5	0.10	1.0–1.5	0.25	0.1	0.1	0.1	0.05	0.05	0.05–0.25	0.05	0.15
BSL155		3.8–4.5	0.10	1.0–1.5	0.25	0.1	0.1	0.1	0.05	0.05	0.05–0.25	0.05	0.15
BSL169	(b)	0.10	0.50–0.75	6.5–7.5	0.20	0.10	0.05	0.10	0.05	0.05	0.10–0.20	0.05	0.15
BSL173	(b)(d)	0.20	0.25–0.45	6.5–7.5	0.20	0.10	—	0.10	—	—	0.04–0.25	0.05	0.15
BSL174	(b)(d)	0.20	0.25–0.45	6.5–7.5	0.20	0.10	—	0.10	—	—	0.04–0.25	0.05	0.15
DTD716B		0.1	0.3–0.8	3.5–6.0	0.6	0.5	0.1	0.1	0.05	0.05	0.25	—	—
DTD722B		0.1	0.3–0.8	3.5–6.0	0.6	0.5	0.1	0.1	0.05	0.05	0.25	—	—
DTD727B		0.1	0.3–0.8	3.5–6.0	0.6	0.5	0.1	0.1	0.05	0.05	0.25	—	—
DTD735B		0.1	0.3–0.8	3.5–6.0	0.6	0.5	0.1	0.1	0.05	0.05	0.25	—	—
DTD5008B	(c)	0.1	0.5–0.75	0.25	0.5	0.1	0.1	4.8–5.7	0.05	0.05	0.25	—	—
DTD5018A		0.2	7.4–7.9	0.25	0.35	0.1–0.3	0.10	0.9–1.4	0.05	0.05	0.25	—	—

Notes: Single figures are maxima.
(a) Zr 0.10–0.30 Co 0.10–0.30
    Sb 0.10–0.30 Ti + Zr 0.50 max.
(b) Be 0.07 max.
(c) Cr 0.4–0.6.
(d) Specific modifying elements may be added in sufficient quantities to provide the necessary casting material qualities.
Refer to BS4L101 for details.

Table 2.7 (Continued)

Aerospace	BS1490	ISO	EN	France	Germany	Italy	USA AA	USA SAE
BS2L99	LM25	AlSi7Mg	42 000	A–S7G	G–AlSi7Mg	UNI 7257	A356	336
BSL119				A–U5NKZr				
BSL154							295	
BSL155							295	
BSL169				A–S7GO6	G–AlSi7Mg	UNI 8392	A357	
BSL173	LM25	AlSi7Mg	42 000	A–S7G	G–AlSi7Mg	UNI 7257	A356	336
BSL174	LM25	AlSi7Mg	42 000	A–S7G	G–AlSi7Mg	UNI 7257	A356	336
DTD716B		AlSi5Mg		A–S4G	G–AlSi5Mg (235)	UNI 3054		
DTD722B		AlSi5Mg			G–AlSi5Mg (235)	UNI 3054		
DTD727B		AlSi5Mg			G–AlSi5Mg (235)	UNI 3054		
DTD735B		AlSi5Mg			G–AlSi5Mg (235)	UNI 3054		
DTD5008B	LM31	AlZn5Mg	71 000	A–Z5G		UNI 3602	712	310
DTD5018A		AlMg8Zn			G–AlMg9 (349)	UNI 3057	535	

Notes:

1. The foreign specifications given above are only an indication of the alloy type.
2. BSL and DTD specifications are subject to BS4L101 and other inspection requirements.

Table 2.8 Casting characteristics, etc.

Alloy	Sand cast	Gravity die	L.P. die	H.P. die	Fluidity	Hot tear resist.	Pressure tightness	Machinability	Corrosion resistance	Density (g/cm³)	Coeff. of expansion per °C × 10⁻⁶	Thermal conduct. at 25°C (W/mK)	Electrical conduct. at 20°C (% IACS)
LM0	F	F	F	F	F	F	F	F	E	2.70	24	209	57
LM2	G*	G	G	F	G	E	E	F	F	2.74	20	100	26
LM4	E	E	G	G*	G	G	E	G	F	2.75	21	121	32
LM5	F	F	F*	F	F	F	P	E	E	2.65	23	138	31
LM6	E	E	E	G	E	E	E	P	E	2.68	20	142	37
LM9	G	G	E	*	G	E	G	F	G	2.94	22	147	38
LM12	F	G	P*	*	F	G	G	E	P	2.70	22	130	33
LM13	G	G	G	*	G	E	F	F	G	2.70	19	117	29
LM16	G	G	G	*	G	E	G	G	G	2.68	23	142	36
LM20	E*	E	E	E	E	E	E	P	G	2.81	20	155	37
LM21	G	G	G	*	G	G	G	G	F	2.77	21	121	32
LM22	G*	G	G	*	G	G	G	G	F	2.79	21	121	32
LM24	F*	F	F	E	G	E	E	G	E	2.68	21	96	24
LM25	E	E	E	G*	G	E	E	G	F	2.76	22	151	39
LM26	G	G	G*	*	G	E	F	G	G	2.75	21	105	26
LM27	E	E	G*	G*	G	E	E	G	G	2.68	21	155	27
LM28	P	F	P	*	F	G	F	P	G	2.65	18	134	—
LM29	P	F	P	*	F	G	F	P	G	2.73	16	126	—
M30	*	F	F	G	G	F	F	P	F	2.81	18	134	20
LM31	G	F	F	F	F	F	F	G	E	2.81	24	147	35

Notes: * Not normally used in this form.
E = excellent   G = good   F = fair   P = poor.

L.P. die = Low-pressure diecast.
H.P. die = High-pressure diecast.
Coeff. of linear expansion relates to temperatures in the range 20–100°C.
Thermal conductivity will vary with the condition of the casting.
Electrical conductivity is quoted in percentage of IACS (International Annealed Copper Standard Units).

Table 2.9   Uses and general remarks

LM0	Electrical, food and chemical plant.
LM2	Pressure diecast components.
LM4	Popular alloy for sand, gravity and low-pressure diecast engineering components; cylinder heads, manifolds, engine mounts, gear boxes, etc.
LM5	Sand and chill castings where corrosion resistance is needed; marine, food, chemical plant.
LM6	Sand and chill castings, particularly for thin-walled intricate castings.
LM9	Has the casting characteristics of LM6 but can be heat treated to higher strength.
LM12	Sand and chill cast; used where a good machined surface is needed, hydraulic brake castings.
LM13	Sand or chill; used for pistons.
LM16	Intricate sand or chill castings, good strength after heat treatment, cylinder heads and blocks, etc.
LM20	Mainly used for pressure diecastings, but can be sand cast. Better corrosion resistance than LM24.
LM21	Sand or chill cast, similar to LM4 but higher proof stress.
LM22	Usually chill cast and solution treated. Good shock resistance, used for heavy duty castings.
LM24	Popular pressure diecasting alloy.
LM25	Popular sand and chill casting alloy; good for thin section, leak-tight castings, good corrosion resistance; cylinder heads, blocks, wheels, etc.
LM26	Pistons for petrol and diesel engines.
LM27	Versatile sand and chill casting alloy for thick or thin sections, slightly superior to LM4, used for the same applications.
LM28	High performance pistons, best cast in permanent moulds to optimise structure.
LM29	As LM28 but lower coefficient of expansion. Like LM28, needs special foundry technique.
LM30	Used for pressure diecast, unlined cylinder blocks; low expansion and good wear resistance.
LM31	Used for large sand castings, good shock resistance.

Table 2.10   Colour codes for BS1490 ingots. This colour code chart has been incorporated into BS 1490 and provides a common scheme for marking ingots

	White	Green	Black	Brown	Blue	Yellow	Red
White	LM0						
Green		LM12					
Black		LM30	LM5				
Brown	LM28	LM29					
Blue	LM31	LM27			LM26		
Yellow		LM20	LM13	LM9	LM25	LM6	
Red	LM2	LM21	LM16	LM22	LM24		LM4

A full listing of standards for aluminium alloys, "Comparison of National Standards for Aluminium Casting Alloys", was published in June 1996 by OEA (Organisation of European Aluminium Refiners and Remelters, Broadway House, Calthorpe Road, Five Ways, Birmingham B15 1TN, England). The list covers all European Standards together with Australia, Japan, Korea, South Africa and the USA.

Casting processes

Aluminium alloys can be cast by several processes:

> Sand casting
> Chill casting, i.e. gravity die (permanent mould casting) or low pressure diecasting, in a metal die using sand cores
> Pressure diecasting
> Lost Foam
> Squeeze casting
> Investment casting or plaster moulding precision casting techniques may also be used for aerospace castings

Casting alloys are designed to be cast by one or other of these techniques, although some alloys are suitable for more than one casting method.

The effect of alloying elements

Silicon

Pure aluminium melts at 660.4°C it is not suitable for casting and is only used for electrical applications (where high conductivity is essential), and a few other special applications. Most casting alloys contain silicon as the major alloying element. Silicon forms a eutectic with aluminium at 11.7% Si, 577°C (Fig. 2.2). Silicon additions improve casting characteristics by improving fluidity, feeding and hot tear resistance. The silicon-rich phase is hard, so the hardness of the alloy is increased with Si content but ductility and machinability are reduced.

Typical silicon levels of popular casting alloys are:

Alloy	Si content	BS alloy	Typical freezing range (°C)
Low silicon	4–6%	LM4	625–525
Medium silicon	7.5–9.5%	LM25	615–550
Eutectic alloys	10–13%	LM6	575–565
Special hypereutectic alloys	> 16%	LM30	650–505

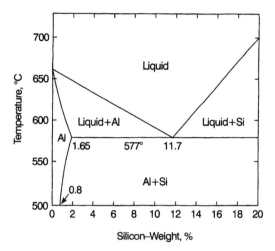

Figure 2.2   *The Al–Si phase diagram.*

The eutectic alloys have the highest fluidity for a given casting temperature and having a short freezing range, they solidify with primary shrinkage. They are good for thin section castings. Where higher strength is needed, the lower silicon alloys are used. The hypereutectic alloys are difficult to machine, they are used for wear-resistant applications such as pistons.

Copper

Improves strength, hardness, machinability and thermal conductivity. Heat treatment is most effective with 4–6% Cu alloys. Copper decreases castability and hot tear resistance together with corrosion resistance.

Magnesium

Small additions of 0.25–0.5% Mg allow Al–Si alloys to be hardened by heat treatment, improving mechanical properties through the precipitation of Mg_2Si in a finely dispersed form. Their proof stress can be almost doubled. Mg is used at levels of around 1% in high silicon piston alloys. Higher levels still, around 3–6% Mg, are used in low silicon alloys to improve the anodising characteristics and give a bright surface finish for decorative components. Magnesium content is kept low in pressure diecasting alloys to avoid embrittlement. The presence of magnesium increases the oxidation losses of liquid aluminium.

Iron

Levels of 0.9–1.0% Fe are used in pressure diecasting alloys to prevent die sticking. High Fe contents decrease ductility, shock resistance and machinability. Castability is decreased by Fe due to the formation of sludge phases with manganese and chromium etc., so alloys for processes other than pressure diecasting are limited to less than 0.8% Fe.

Manganese

Improves casting soundness at levels over 0.5%. Mn controls the intermetallic form of iron in the alloy, leading to improved ductility and shrinkage characteristics.

Nickel

When combined with copper, enhances strength and hardness at elevated temperature.

Zinc

When combined with copper and magnesium, heat treatment and natural ageing characteristics are improved. The fluidity is increased but shrinkage problems may occur.

Lead

Improves machinability at levels over 0.1%.

Titanium

Refines the grain structure when combined with boron.

Phosphorus

Refines the primary Si phase in hypereutectic alloys. In hypoeutectic alloys, low levels of phosphorus coarsen the eutectic structure and reduce the effect of Na and Sr eutectic modifiers.

Strontium

Levels of 0.008–0.04% Sr modify the Al–Si eutectic structure.

Sodium

Used to modify the eutectic structure.

Lithium

While lithium up to 3% may be used to improve the properties of wrought aluminium alloys, it has a generally harmful effect on casting properties by reducing the effectiveness of Na or Sr modifiers at levels above 0.5%. At even lower levels, above 0.01%, porosity problems are experienced. It is recommended that Li levels below 0.003% are used for casting alloys.

Heat treatment of aluminium alloys

Heat treatment designation for casting alloys

Suffixes are given to alloy designations to indicate the heat treatment condition, for example LM25-TF is a casting conforming to the BS 1490 chemical specification, supplied in the fully heat-treated condition, being solution treated and artificially aged, Table 2.11.

Table 2.11   Heat treatment suffixes

Condition	Suffix		
	UK	USA	EN
As cast	M	F	F
Annealed	TS	O	O
Controlled cooling from casting and naturally aged	–	–	T1
Solution heat treated and naturally aged where applicable	TB	T4	T4
Solution heat treated and stabilised	TB7	–	–
Controlled cooling from casting and artificially aged or overaged	TE	T5	T5
Solution heat treated and fully artificially aged	TF	T6	T6
Solution heat treated and artificially underaged	–	–	T64
Solution heat treated and artificially overaged	–	T7	T7
Solution heat treated, artificially aged and stabilised	TF7	–	–

Many castings are used in the as-cast condition, but certain applications require higher mechanical properties than the as-cast material. The proof stress of castings of alloy Al–Si7 Mg (LM25), for example, can be more than doubled by full heat treatment. Table 2.6 shows the effect of heat treatment on common casting alloys. For sand, gravity and low pressure diecastings, all treatments are possible, though not all are standardised. Pressure diecastings are not solution treated and quenched in the same way as sand and permanent mould castings. Most pressure diecastings contain bubbles of air or gas from the die lubricant trapped in the metal as it is injected into the die. If the castings are solution treated, the trapped gas bubbles expand and may cause pimples and distortion on the surface of the casting. Pressure diecastings made using special processes such as vacuum diecasting or squeeze casting contain less gas and may be solution treated.

All diecastings may be quenched from the die, precipitation treated and stress relieved without suffering harmful effects.

Heat treatment furnaces

Some heat treatments are carried out close to the melting point of the castings so accurate temperature control is needed. Forced air circulation furnaces are used to ensure that the temperature of all parts of the furnace is constant.

Stress relieving and annealing (TS condition)

Castings having changes of section, or having complex shape are likely to develop internal stresses in the mould or die because of differential cooling. The internal stresses may be released when the casting is machined, causing dimensional changes. To remove internal stresses, castings are heated to a temperature of 200°C for 5 hours followed by slow cooling in the furnace.

Solution treatment (TB condition)

Castings are heated at temperatures just below the melting temperature for a long time to take the alloying constituents into solid solution. The castings are then rapidly cooled by quenching to room temperature to retain the elements in solution. Water (often hot) or special quenchants are used. The quench tanks are placed close to the furnace to ensure rapid cooling.

Solution treated and stabilised (TB7 condition)

Solution treatment is followed by stress relief annealing.

Precipitation treatment (TE condition)

Controlled precipitation of alloying constituents is promoted by heating the casting to a temperature between 150°C and 200°C for a suitable time. Strength and hardness are increased. With chill castings (made in dies), it is possible to obtain some increase in strength of as-cast components by precipitation treatment since the rapid cooling in the die retains some of the alloying constituents in solution. The time of treatment is important, since too long a time at temperature will result in a reduction in the mechanical properties. Each alloy has an optimum heat treatment cycle, examples are given in Table 2.12.

Table 2.12   Examples of heat treatment times and temperatures

Alloy and condition	Solution treatment			Precipitation treatment	
	Time (hrs)	Temp(°C)	Quench	Time(hrs)	Temp(°C)
Al–Si5Cu3 LM4-TE	6–16	505–520	Hot water	6–18	150–170
Al-Si7 Mg LM25-TE				8–12	155–175
LM25-TB7	4–12	525–545	Hot water	2–4	250
LM25-TF	4–12	525–545	Hot water	8–12	155–175

Note: Times do not include time to reach temperature.
TE Precipitation treated; TF Solution and precipitation treated; TB7 Solution treated and stabilised

Solution treated, quenched, precipitation treated and stabilised (TF condition)

Castings used at elevated temperature, such as pistons, benefit from stabilisation treatment at 200–250°C following precipitation treatment. Some reduction in mechanical properties occurs.

Age hardening and artificial ageing

Some casting alloys improve their strength and hardness while standing at room temperature. The process may take several weeks but can be speeded up by heating above room temperature.

For more details, refer to specialist heat treatment manuals such as *British and European Aluminium Casting Alloys, their Properties and Characteristics* published by The Association of Light Alloy Refiners, Birmingham, England, from which this summary has been taken.

Chapter 3
Melting aluminium alloys

Introduction

The successful casting of aluminium alloys requires attention to a number of special factors.

Oxidation

Molten aluminium and its alloys immediately oxidise when exposed to air forming a skin of oxide. In pure aluminium this is Al_2O_3 but the presence of Mg in the alloy can cause the oxide to form as $MgO.Al_2O_3$ (spinel). The oxide skin has a protective effect, preventing catastrophic oxidation of the melt (which occurs when magnesium is melted) but it causes problems during melting and also during casting. An oxide film can form even as the metal is filling the mould and can give rise to entrained oxide in the casting harming the physical properties of the casting and possibly causing leaking castings. Professor Campbell has drawn attention to the harmful effects of entrained oxide films in aluminium alloy castings, Table 3.1.

Oxide inclusions in aluminium alloys are of Al_2O_3 or $MgO.Al_2O_3$ which have a density only 5% less than that of liquid aluminium so flotation of oxide inclusions takes place slowly. For inclusion-free castings it is advisable to use metal filters to clean the metal as it enters the mould (see Chapter 8). Fluxes are used during melting to protect the metal from oxidation and to trap oxides as they float out of the melt.

Table 3.1   Forms of oxide in liquid aluminium alloys

Growth time	Thickness	Type	Description	Possible source
0.01–1 sec.	1 μm	New	Confetti-like fragments	Pour and mould fill
10 sec.–1 min.	10 μm	Old 1	Flexible, extensive films	Transfer ladles
10 min.–1 hr	100 μm	Old 2	Thicker films, less flexible	Melting furnace
10 hr–10 days	1000 μm	Old 3	Rigid lumps and plates	Holding furnaces

(From *Castings*, Campbell, J. (1991), Butterworth-Heinemann, reproduced by permission of the publishers.)

Hydrogen

Molten aluminium readily picks up hydrogen from the atmosphere or from moisture-containing refractories, the solubility of hydrogen in solid aluminium is very low, so that as the alloy freezes hydrogen gas is expelled, causing micro- or macro-porosity in the casting. To achieve high integrity castings, aluminium alloy melts must be degassed before casting.

Structure

The microstructure, and therefore the mechanical properties, of Al–Si alloys can be modified and improved by appropriate metal treatment. "Modifiers" and/or grain refiners are usually added to the alloy before casting.

Feeding

Aluminium alloys shrink on freezing so that castings must be correctly fed to achieve soundness.

To avoid the above problems, great care must be taken at all stages of melting, treatment and casting of aluminium alloys.

Raw materials

Foundries usually purchase pre-alloyed ingots from specialist suppliers who convert miscellaneous scrap into high quality, accurately specified material subject to national standards such as BS1490:1988 or the European CEN Standard (Table 2.1). Scrap metal is carefully sorted by the supplier using spectroscopic analysis and melted in large induction or gas-fired furnaces. There is always a danger of contamination by impurities, particularly iron, and by alloying additions such as magnesium and silicon. The supplier must take great care to ensure that the standards are complied with. Ingots are usually about 5 kg in weight and may be colour coded to avoid danger of mixing (see Table 2.10).

Melting furnaces

A wide range of furnace types is used by aluminium foundries. Small foundries may use lift-out crucible furnaces in which the metal is melted and treated in a crucible which is then lifted out of the furnace for pouring. Large foundries usually melt aluminium alloy ingot and foundry returns in a bulk melting furnace, then transfer the metal to smaller holding furnaces near to the casting area. Degassing and metal treatment are usually carried out in the transfer ladle. The bulk melting furnaces can be coreless electric induction furnaces or, more commonly, gas-fired reverberatory or shaft furnaces. The tilting crucible furnace, which may be electric or gas, is also popular as a bulk melter. Holding furnaces may be electric or gas.

Coreless induction furnaces

Medium frequency induction furnaces are efficient, clean and rapid melting units for aluminium, Fig. 3.1. Aluminium induction furnaces usually range from 500 kg to 2 tonnes capacity and operate at frequencies of 250–1000 Hz.

For example, in one installation, two 1.5 tonne aluminium capacity steel shell tilters are powered by a 1250 kW, 250 Hz power supply with a changeover switch which allows alternate furnaces to be melted. When ready, the furnaces are hydraulically tilted into a transfer ladle or by launder to adjacent holding furnaces. 1.5 tonnes can be melted in 40 minutes. While induction furnaces are excellent melting units, they are not efficient holders. When used for melting, it is advisable to transfer the molten metal to an efficient holding furnace as soon as it has reached the required temperature.

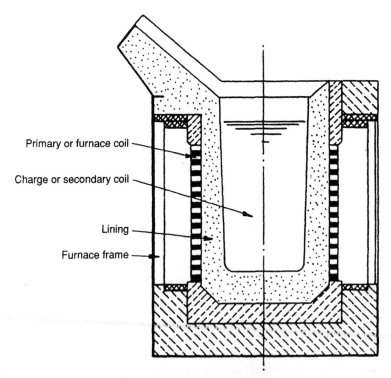

Primary or furnace coil

Charge or secondary coil

Lining

Furnace frame

Figure 3.1 *Section through a coreless induction furnace.*

Induction furnaces are energy efficient melters. Energy consumption for melting is affected by the density of the charge and the melting practice used. Batch melting is less efficient than using a molten heel, a 50% molten heel being most efficient. Energy consumptions vary from 540 kWh/tonne (2 GJ/tonne) for a high bulk density charge (small scrap and ingot) to 600 kWh/tonne (2.2 GJ/tonne) if lower density scrap (such as pressure diecasting runners and ingot) is melted. While energy consumption is low, costs for melting may be higher than for gas-fired furnaces because of the generally high cost of electricity as a source of heat.

The stirring effect of the induction power can be advantageous since the charge is mixed well but it exposes the molten aluminium to oxidation and the oxide may be drawn into the melt which can be harmful and lead to high melting losses. The stirring effect causes fluxes to be entrained in the melt, so it is usual to melt without flux cover, then to switch off the current before adding the flux. Sufficient time must be allowed for the oxides to float out before transferring the metal.

The linings are usually a dry alumina refractory, vibrated around a steel former according to the supplier's instructions and heated at 80–100°C/hr to 750°C then held for 1–4 hrs depending on the size of the furnace and cooled naturally before removing the former.

Dross build-up on the linings can reduce furnace efficiency and contribute to lining failure. Dross should be scraped from the walls at the end of each melt cycle while the furnace is hot. Once a week, the furnace should be allowed to cool and any remaining dross carefully removed using chisels.

Reverberatory furnaces

Reverberatory furnaces have gas or oil burners firing within a refractory hood above the metal bath, Fig. 3.2. The burner flame is deflected from the roof onto the hearth. Waste heat is used to preheat the charge as it descends down the flue. They are used as batch melters. They are simple and have relatively low capital cost which makes them attractive for bulk melting of ingots and foundry returns. They are produced in a variety of configurations

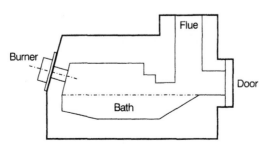

Figure 3.2 *Cross-section of a Sklenar reverberatory furnace. (Courtesy Ramsell Furnaces Ltd, Droitwich.)*

such as fixed or tilting, rectangular or cylindrical with melting capacities from 200 to 1300 kg/hr. Large reverberatory furnaces give rapid melting and can handle bulky charge material, but the direct contact between flame and charge may lead to high metal losses, gas pick-up and considerable oxide contamination. Temperature control can also be difficult. This type of furnace is being used less because of its relatively low thermal efficiency of around 1100 kWh/tonne.

Shaft furnaces

Higher thermal efficiency can be achieved in a tower or shaft furnace, Fig. 3.3. These furnaces are both melting and holding units. They consist of three chambers, the first is a preheating area charged with a mixture of foundry returns and ingot by a skip charging machine. Waste heat from the melting and holding burners heats the charge removing moisture and oil before melting takes place. The preheated charge then enters the gas-fired melting zone and the liquid aluminium runs down into the holding bath. Here, the temperature is accurately controlled within ±5°C. Typical shaft furnaces range in size from a holding capacity of 1000 kg and a melting rate of 1000 kg/hour to over 3000 kg holding and 3000 kg/hr melting capacity. Shaft furnaces of much larger capacity are also available.

Molten aluminium is discharged to a transfer ladle or launder either by hydraulically tilting the holding bath or by a tap-out system.

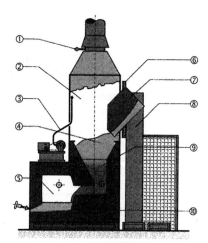

Figure 3.3 *Gas-fired shaft furnace. The STRIKO ETAmax system. (Courtesy STRIKO UK Ltd.)*

1 Waste gas temperature control	6 Charging door
2 Waste gas hood	7 Charging car
3 Baffle	8 Charging unit
4 Preheating zone	9 Shaft/melting zone
5 Holding chamber	10 Furnace body

Energy consumption of 580–640 kWh/tonne (2–2.3 GJ/tonne) can be achieved with melting losses of 1–1.5% when melting 50/50–ingot/foundry returns. Operating the furnace below rated capacity has a significant effect on energy consumption. A furnace working at 50% of its rated throughout may use almost twice as much energy per tonne (1070 kWh/tonne, 3.8 GJ/tonne).

Typical metal loss in a shaft furnace melting about 50% ingot, 50% foundry returns is 1–1.2%. Refractory life is high, with relining needed every 3 or 4 years. Cleaning once per shift is necessary to avoid corundum build-up.

Crucible furnaces

Crucible furnaces, Fig. 3.4, are widely used as melters, melter/holders and holders. Crucible furnaces are:

    simple and robust
    widely available in a range of sizes
    either fixed or tilting
    suitable for heating by different fuels
    capable of low melting losses
    relatively inexpensive

Alloy changes are readily carried out and both degassing and metal treatment can be done in the crucible before it is removed for casting.

Crucible furnaces fall into three main types:

Lift-out   The crucible is removed from the furnace for pouring
Tilting   The furnace body containing the crucible is tilted to pour the molten metal
Bale-out   The molten metal is ladled out

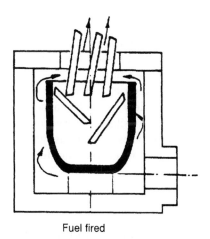

Fuel fired

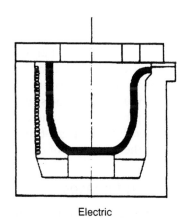

Electric

Figure 3.4 *Schematic cross-sections of tilting crucible furnaces: (a) fuel fired, (b) electric. (Courtesy Ramsell Furnaces Ltd, Droitwich.)*

Crucibles may be either clay graphite (clay bonded) or silicon carbide (carbon or resin bonded).

Clay graphite crucibles consist of special graphites with clay as the bonding agent. The clay forms a ceramic bond, some silicon carbide may be added to improve resistance to thermal shock. The graphite provides thermal and electrical conductivity and resistance to wetting by molten metal or salts. The crucible is coated with a glaze which prevents oxidation of the graphite.

Silicon carbide crucibles consist of SiC and special graphites. They are carbon bonded using pitch, tar or a resin. The crucibles are glazed to ensure high resistance to oxidation. While silicon carbide crucibles are more expensive than clay graphite, their life is longer.

Crucible life has increased with advances in manufacturing methods, and in furnaces used mainly for holding, crucible lives of twelve months or more are possible with careful use. The main points to pay attention to are:

> Avoid mechanical shock
> Use padded tools for transport
> Do not roll the crucible on its bottom edge or side
> Avoid damage to the protective glaze
> Crucibles can absorb moisture which can give rise to spalling of the glaze when heating up
> Store in a dry place, not on a damp floor

The crucible should always be preheated before charging, following manufacturer's instructions. It should be charged as soon as it has reached red heat (about 800°C). The crucible wall must be cleaned immediately after emptying to remove slag or dross. If not removed immediately the slag or dross will harden and be difficult to remove.

The tilting crucible furnace (which may be gas-fired or electric resistance heated) remains popular for batches of aluminium up to 700 kg. The crucible tilts to discharge metal into casting ladles. Thermal efficiency is not as high as some other melting furnaces since it is difficult to make use of the heat in the products of combustion. They are relatively inexpensive and since the flames are not in contact with the molten metal, metal losses are low and melt quality high and alloy changes are readily carried out.

Electrically heated crucible furnaces having electric resistance wire elements or silicon carbide rod-type elements are widely used. The absence of combustion products in the furnace reduces the possibility of hydrogen pick-up by the metal and avoids the considerable heat loss to exhaust gases suffered by oil- or gas-fired units. These advantages offset to some degree the higher unit cost of electricity.

Holding furnaces

Melting aluminium in a bulk melting furnace exposes the liquid metal to turbulence and oxidation. The low density of aluminium retards the "float-out" of oxide inclusions, and it is desirable to allow the liquid alloy to stand

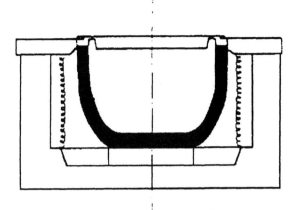

Figure 3.5 *Electrically heated crucible furnace. (Courtesy Ramsell Furnaces Ltd, Droitwich.)*

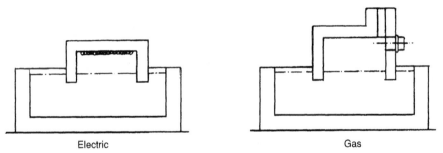

Electric   Gas

Figure 3.6 *Radiant roof holding furnace. (Courtesy Ramsell Furnaces Ltd, Droitwich.)*

in tranquil conditions to allow the non-metallics to float out before transferring to the casting ladle. A holding furnace is used. They are frequently resistance-heated crucible furnaces, Fig. 3.5, or radiant-roof bath furnaces, Fig. 3.6, in which high insulation allows low holding power to be achieved. Capacities are typically 250–1000 kg, although much larger holding furnaces are possible.

The bale-out and charge wells, which are fitted with insulated covers, are generally small to reduce heat losses while the covers are off. The wells are separated from the main bath by refractory walls with connecting holes at the bottom to allow clean metal to pass from one area to another. With good thermal insulation, close temperature control is possible with very low energy costs.

In the Cosworth casting process, and other similar processes, molten aluminium from the bulk melting furnace is transferred by transfer pump or by tilting to a large electric radiant-heated holding furnace of sufficient capacity to allow the liquid metal to stand under tranquil conditions for 1 or 2 hours to allow time for the oxides to float out. An electromagnetic pump, drawing from the middle of the bath, fills the sand moulds with inclusion-free metal. The holding furnace is automatically refilled from the melting furnace.

In less critical applications, such as pressure diecasting, or in foundries where inclusion control is accomplished by filtration of the metal in the mould, the holding furnace need not be so large and may be designed to allow alloy adjustment, temperature control and some metal treatment before transfer to the casting ladle.

In pressure and gravity diecasting foundries, it is convenient to have a holding furnace adjacent to the diecasting machine in which metal is held at the correct temperature and from which it may be baled out to fill the die.

Dosing furnaces

Pressure displacement dosing furnaces are designed to hold aluminium at temperature at the casting station and to automatically meter accurate charges of metal to the die by pressure displacement through a refractory riser tube. Accuracy of pour is within ±1.5%. They can be used to feed pressure diecasting machines and gravity-die carousels.

Corundum growth

Corundum (Al_2O_3) is formed when aluminium comes into contact with silica in the furnace lining. Corundum growth is well known in the aluminium melting industry. It is a composite of alumina and metal which grows on the refractory wall above the metal level in holding furnaces. The growths are extremely hard, smooth and initially hemispherical. They are difficult to remove and when viewed in the hot furnace are generally grey or black ranging in size from a few millimetres diameter to tens of centimetres. The growth direction is generally away from the metal line, upwards towards the roof of the furnace in a mushroom shape. Corundum growth not only reduces capacity of the furnace but it reduces the thermal efficiency and causes damage to the furnace lining through refractory expansion. A significant amount of aluminium metal may also be lost from the furnace charge.

To avoid serious corundum growth, regular inspection of the furnaces must be carried out and growths removed while they are small. The furnace refractories should be resistant to metal attack, by having a high bauxite content and low free silica content. Refractories should be non-wetting and of low porosity to avoid corundum nucleation. High temperature, oxidising furnace atmospheres and the presence of unburned hydrocarbons should be avoided. Daily cleaning of the furnace refractories with a suitable flux is advisable (see Chapter 4).

Choice of melting unit

The number of alloys required by the foundry is a major factor in deciding the type of melting furnace used. A sand foundry may use several different alloys each day. In this case, tilting crucible furnaces may be the best solution even though they may not be the most fuel or labour efficient. A pressure diecasting foundry, on the other hand, may melt a single alloy only so a bulk-melting tower furnace or induction furnace supplying small holding furnaces at each diecasting machine is likely to be the lowest cost solution.

Most gravity diecasting foundries have some alloys which do not warrant bulk melting, so in addition to a bulk melter the foundries usually have some smaller melting furnaces, often of the crucible type.

Chapter 4
Fluxes

Introduction

Chemical fluxes for aluminium have a number of functions:

Covering fluxes which form a molten layer to protect the melt from oxidation and hydrogen pick-up

Drossing-off fluxes which agglomerate the oxides allowing easy removal from the surface of the melt

Cleaning fluxes which remove non-metallics from the melt by trapping the oxide particles as they float out

Fluxes which "modify" the alloy, by introducing sodium, improving its microstructure

Exothermic fluxes which ensure that aluminium liquid trapped in the dross layer is returned to the melt

Fluxes for reclaiming swarf, skimmings and turnings, giving a high metal yield

Fluxes for the removal of oxide build-up from furnace walls

Some fluxes combine several of these functions. The Foseco range of COVERAL fluxes is designed to be used on a range of alloys in a variety of melting units. For many years fluxes have been supplied in powder form. Table 4.1 lists the various COVERAL fluxes and their uses. A recent Foseco development is the range of COVERAL GR granulated fluxes, Table 4.2 These have significant environmental and technical advantages over the traditional powder fluxes and are rapidly replacing them.

In general, the lower the melting point of the cover flux, the more efficient its use. Fluxes for aluminium contain chemicals such as chlorides and fluorides which may give rise to potentially harmful fumes in use on molten metal. Operators must avoid inhalation of the fumes or dust. Used flux must be disposed with care, referring to the local authority or a specialist disposal company for instructions.

Application of COVERAL powder fluxes

Covering and protecting during melting

Aluminium alloys containing up to 2% Mg are usually treated with dry fluxes in crucible and induction melting and with liquid fluxes in reverberatory, shaft, rotary and large electric furnace melting. The required flux is selected from Table 4.1. Sufficient COVERAL to form a cover (usually about 0.5–1% of the metal weight) is added, preferably in two stages, half early in the melting procedure and the remainder as soon as the charge is fully molten. The cover should be kept intact if possible until the melt is ready for degassing and grain refining.

Most fluxes contain sodium and it is possible for the metal to pick up as much as 0.001% Na from them. For most aluminium alloys the sodium has no effect or is beneficial, but alloys containing more than 2% Mg may become brittle with even trace amounts of sodium, so they are treated with one of the sodium-free fluxes shown in Table 4.1. In the case of COVERAL 65, approximately 0.5% of the product is put onto the solid charge and a further 2% sprinkled evenly over the surface when the alloy is fully molten. When the flux becomes pasty or liquid at about 750°C, the flux is worked well into the melt with a bell plunger for about 3 minutes.

Drossing-off before pouring

The function of a drossing-off flux is to absorb oxides and non-metallic material, cleansing the metal and forming a good metal-free dross which can easily be removed.

In crucible furnaces, when drossing-off is carried out, the crucible sides are scraped and the required quantity of the selected COVERAL (250 g is normally enough for the lift-out or bale-out furnace) is sprinkled onto the metal surface along with the existing flux cover and mixed into the surface of the melt until a red-glowing dross is obtained. This is exceptionally free of metal and can be removed with a perforated skimmer.

In reverberatory and shaft furnaces, the quantity of flux needed will depend on the cleanliness of the charge material and on the surface area of the metal. As a guide, it is recommended that an application of $1–2 \, kg/m^2$ will suffice. The behaviour of the flux will indicate whether the dosage needs to be reduced of increased in future applications.

When the melt is ready for drossing-off, the flux is spread over the metal surface, allowed to stand for a few minutes until fused and then rabbled into the dross for several minutes with a skimmer. For best results the melt should preferably be above 700°C although fluxes will function well below 650°C. The furnace is then closed and the flame turned on for 10 minutes. This helps to activate the flux, heating the dross and giving good metal separation. The dross is then pulled to the door, allowed to drain and transferred to a dross bogie. If the dross in the bogie is raked, further metal will collect in the bottom.

Reclamation of swarf, skimmings and turnings

A heel of metal is melted using heavy scrap or ingot and a quantity of COVERAL 48 flux is added to form a fluid cover. The amount of COVERAL

Table 4.1 COVERAL fluxes for melting aluminium alloys

Grade	Flux type	Melt pt °C	Dross type	Melting unit	Alloy type	Remarks	Method of application
5F	Covering Drossing	620	Dry	Large, rev., shaft rot, tilter etc.	All low Mg alloys	Also prevents furnace wall build-up, usually used in smelters	Add half early, rest at final melt-down. Form a complete cover. Use about 0.5–1% of metal weight or 1 kg/m² of melt area
11	Covering Drossing	–	Dry	Crucible, electric	ditto	Used for ingot and clean scrap melted for sand and diecastings	Add 0.5–1% early and maintain cover. Remove after grain refining/degassing
29A	Modifying	780	Liquid	Crucible, bale-out	9–13% Si	Used at 760–800°C	Melt under COVERAL 11, heat to 800°C, grain refine/degas skim, add 1% COVERAL 29A, work in, leave 5 min., dross off with COVERAL 11
36A	Modifying	670	Liquid	Crucible, bale-out	7–13% Si	Low temp. version of 29A	Melt under COVERAL 11, degas and skim, add 2% COVERAL 36A at around 750°C and stir in, leave 5–10 min. and dross-off with COVERAL 11
48	Refining	590	Liquid	All	Low-Mg alloy scrap	Reclamation of turnings etc.	Melt heel of solid metal, add 1–5% COVERAL 48 to cover, add further flux with swarf additions
65	Covering	500	Pasty	Crucible	3–10% Mg alloys	Sodium free	Add 0.5% with the charge, add a further 2% when molten, stir in at 750°C. Dross-off after grain refinement and degassing
66	Covering	–	Dry	Crucible, bale-out	1–10% Mg alloys	Sodium free	Add about 0.5% at early stage of melting, a further 0.5% when melting is complete. Dross-off after degassing and grain-refinement
75	Covering Drossing	–	Dry	Holding	Low-Mg alloys	Wide temp. range flux for pressure diecasters	Can be used at 600–800°C, scatter 0.25–0.5% on metal surface and rabble gently until exotherm develops. Push aside remove before taking ladles
88	Furnace cleaning	–	–	Reverb, rotary, transfer ladle	–	To remove oxide build-up formed on furnace walls	Preheat empty furnace to 800–850°C, spray walls using Foseco flux gun, reheat for 15min, then scrape clean. Use once per week
2011	Covering Drossing	–	Liquid	Crucible, electric resist. and induct.	Low-Mg	Low fume covering and dressing flux	Use 0.5–1%, form cover as early as possible and maintain intact. After grain refining and degassing, add further flux, rabble into surface until exotherm, leave 2–3 min., skim

Table 4.2 Granulated COVERAL fluxes for aluminium alloys

Grade	Type	Melting unit	Alloy type	Purpose	Method of application
GR2220	Drossing Cleaning	Bulk melting Transport ladles	Sodium tolerant alloys	Low metal temperature (<680°C) reduces corundum build-up. Suitable for pressure diecasters	Add 0.125–0.25% early, rabble into melt till exotherm develops. Push flux layer aside before pouring into ladle. Regular use keeps furnace walls clean
GR2510	Mild, exothermic drossing	Bulk and smaller furnaces	Compatible with Sr modified alloys	For metal temperatures 650–730°C	Granular equivalent of COVERAL 11. Add 0.125–0.25% early and maintain cover
GR2516	Mild, exothermic drossing	Smaller furnaces	Sodium tolerant alloys	For metal temperatures above 730°C	Add sufficient to form layer, usually 0.125–0.25% of metal. Maintain cover till ready for degassing and grain refining. Dross-off.
GR6511	Sodium free Drossing	Crucible, bale-out	Alloys with 1–10% Mg	Forms protective layer on melt, removes oxides of Mg and Al	Melt under covering flux, degas and skim, add 0.5–1.0%. GR2715, work in for 3 min, leave 5–10 min, dross off with COVERAL 11 or GR2516 and skim
GR2715	Sodium modifying	Crucible, bale-out	Alloys with 7–13% Si	Modifying metallurgical structure to increase ductility for metal temperatures between 700 and 780°C	Best used with Foseco Metal Treatment Station
GR2815	Grain refining	Crucible, bale-out	All alloys	Grain refinement. For metal temperatures >680°C	

used depends on the degree of dirtiness and oxidation of the scrap and will vary between 1 and 5%. The swarf, turnings etc. are fed through the flux cover a little at a time, adding more flux as required to keep the cover in a fluid condition. The temperature of the melt is kept relatively low during this procedure and when charging is complete, the heat is raised to pouring temperature. At this stage the flux may be poured off from rotary or reverberatory furnaces but there is a significant advantage to be obtained by passing more than one melt through the same flux (see Table 4.1).

Modifying aluminium/silicon alloys

The metallurgical structure of alloys containing more than 9% Si is modified to increase ductility (some lower silicon alloys may also benefit from modification). This is done by the introduction of sodium using NAVAC (see Chapter 6) or COVERAL fluxes containing sodium salts (COVERAL 29A, 36A). The metal is melted down under COVERAL 11 and when the required temperature is reached, the melt is degassed and skimmed clean. The modifying flux is then sprinkled evenly over the metal surface and, when pasty or fluid, is worked into the melt for about 3 minutes. After standing quietly for a further 5–10 minutes or until the pouring temperature is reached, the melt is drossed-off with COVERAL 11 and then skimmed.

> COVERAL 29A is used at about 1.5% and a temperature of 790–800°C
>
> COVERAL 36A is used at about 3.0% and a temperature of 740–750°C

Furnace-cleaning flux

Aluminium melting furnace linings become coated with an oxide build-up. COVERAL 88 flux is a strongly exothermic flux which attacks and strips oxide films. The heat generated and the stripping action causes entrapped aluminium to melt and run down to the furnace hearth. Residues on walls are thus loosened and can be removed more easily by scraping tools. The flux is mainly for application to the walls and roof of reverberatory and rotary furnaces. It can also be used for cleaning large transfer ladles, if these can be independently heated. The flux is not recommended for electric furnaces with exposed elements because of the possibility of element attack.

The empty furnace is heated until the lining glows red (800–850°C). Using the Foseco Flux Gun, the walls are sprayed evenly with COVERAL 88. Reheat for 15 minutes then scrape the walls clean. Tap the recovered metal. A 10 tonne furnace will need about 25 kg of flux. Furnaces should be treated weekly to prevent accumulation of build-up.

The flux can be used when making a change of alloy, to prevent contamination of the bath by residues from the preceding charge.

Granular COVERAL fluxes

The formulations of fluxes have not changed for many years, being based on powdered halides including fluorides which are of concern environmentally and which can reduce the life of furnace refractories. Attempts have been made by Foseco to eliminate fluorides completely from the flux formulations, but unfortunately this rendered the flux ineffective. However, it was found that the morphology of the flux had a major effect on its efficiency. By using fluxes in granular form rather than as conventional powders, the effectiveness of the flux can be greatly increased, the handling improved and the undesirable, hazardous emissions can be significantly reduced. The higher cost of granulated fluxes (arising from the additional manufacturing process involved) is compensated by the much reduced quantities needed.

Conventional powder fluxes are normally used at more than 0.25% (by weight of the metal being melted). The granular material is used at only 0.125% by weight so that emissions only half of normal might be expected. In fact, tests have shown fume reduction of more than 85%, Table 4.3.

The tests were carried out on a 250 kg furnace containing 230 kg of A356 alloy (Al–Si7Mg) held at a temperature of 740°C. Powdered flux was added at 0.25% of metal weight, granulated flux at 0.125%. The sampling nozzle was 60 cm above the furnace top.

While emission rates will differ from foundry to foundry, depending on the equipment and practice used, there is no doubt that the move from a powder to a granulated flux significantly improves working conditions,

Table 4.3 Comparison of emissions from a reverberatory furnace without extraction. Concentrations in mg/m^3

	Flux A exothermic drossing flux		Flux B sodium-free drossing flux	
	Powder	Granular	Powder	Granular
Total particulate	3.0	0.53	2.53	0.4
Total Cl	1.04	0.82	1.45	0.71
F	19.0	3.65	14.7	5.05
SO$_x$	15.43	1.18	5.13	2.98

reduces the amount of waste material to be disposed of and reduces attack on furnace refractories.

Table 4.2 lists the Foseco granulated fluxes currently available. The most widely used granular fluxes are:

COVERAL GR2516 (equivalent to COVERAL 11 powder flux)
It is a non-hazardous cleansing and drossing-off flux containing some sodium so should not be used for alloys containing more than 3% Mg where even a slight sodium pick-up must be avoided. Apart from this limitation it is used for most Al alloys. It is particularly recommended for use with crucible furnaces.

COVERAL GR6511 (equivalent to COVERAL 66 power flux)
A sodium-free cleansing and drossing flux for use on alloys sensitive to trace amounts of Na. It is used for all alloys containing Mg in the range 1–10% melted in crucible and bale-out furnaces.

COVERAL GR2220 (equivalent to COVERAL 72 powder flux)
An exothermic drossing flux for use on all sodium tolerant alloys where metal temperature is below 680°C. It is particularly suitable for use with bulk melting shaft furnaces and helps to reduce corundum build-up. Commonly used in pressure diecasting foundries.

COVERAL GR2715 (equivalent to COVERAL 36A power flux)
For the sodium modification of 7–13% Si alloys melted in crucible or bale-out furnaces at low temperatures. Should not be used on strontium modified metal or on hypereutectic alloys (which are modified with phosphorus).

COVERAL GR2815
For the grain refinement of all alloys melted in crucibles or bale-out furnaces. It uses a ratio of 10:1 titanium:boron to nucleate and propagate a fine equiaxed grain growth during solidification. Typical alloys which may be treated are LM4 (Al–Si5Cu3), LM5 (Al–Mg5Si1), LM25 (Al–Si7Mg). COVERAL GR2815 is best used with the Foseco Metal Treatment Station.

Chapter 5

INSURAL refractory for ladles and metal transport

Introduction

It is advantageous in the casting of aluminium alloys to keep metal temperatures as low as possible to avoid the pick-up of hydrogen. INSURAL insulating refractories have been developed specially for aluminium and are widely used for semi-continuous casting and holding processes where gas-free metal and stable temperatures are required. They are also used in aluminium foundries, particularly for ladle liners.

INSURAL refractories have the following properties:

Thermally insulating
Low heat capacity
Non-wetted by molten aluminium
Asbestos free
Durable and resilient structure
Can be machined or sawn if necessary
Controlled density
Simple skull removal

A range of different grades of INSURAL is made. For foundries INSURAL 140 is used (see Table 5.1).

Table 5.1   Properties of INSURAL 140

Density	1.4 g/ml
Thermal conductivity	0.47 W/m/°K
Operating temperature for continuous use	950°C
Maximum operating temperature	1200°C
Modulus of rupture	55 kgf/cm^2

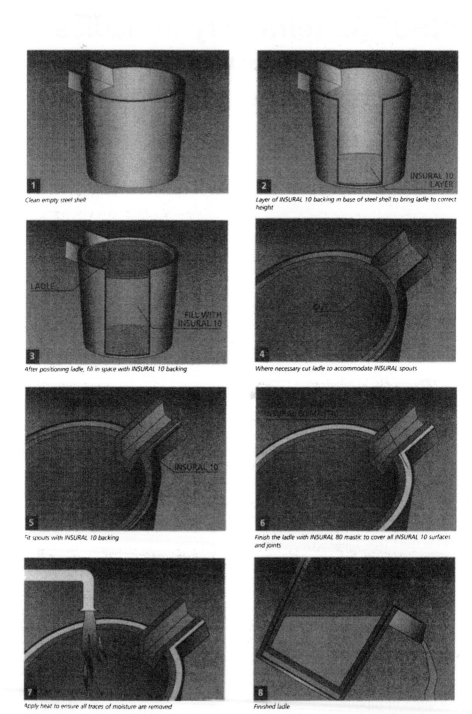

Figure 5.1 *Drawings 1–8. The installation of the INSURAL ATL ladle lining system.*

Ladle liners

Many foundries now use central melting with transfer ladles to move the metal to individual holding furnaces. Frequently, degassing and metal treatment is carried out in the transfer ladle, so the molten metal may be held for up to ten minutes in the ladle. Temperature loss can be a considerable problem and typically the metal must leave the melter at a temperature at least 50°C higher than the holding temperature. Traditionally, dense castable ladle linings have been used. These systems have high conductivity and thermal capacity, so efficient ladle heaters must be used to

(a)

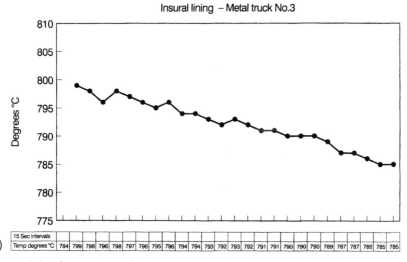

(b)

Figure 5.2 *Comparison of the temperature drop in ladles lined (a) conventionally, (b) with INSURAL ATL.*

ensure effective preheating. Often, however, ladle heating in foundries is quite rudimentary. The use of an uncontrolled gas torch, for example, is extremely inefficient. The use of thermally insulating ladle linings is an economical solution to the problem.

The Foseco INSURAL ATL lining system uses a highly insulating material which is non-wetted by aluminium. Various formulations are available in the INSURAL family of products and INSURAL 140 linings offer a good blend of insulation, thermal cycling and wear resistance. INSURAL 140 preformed ladle linings are designed to be used in conjunction with INSURAL 10 backing material and INSURAL 80 sealing mastic. Together these comprise the INSURAL ATL ladle lining system.

Figure 5.1, drawings 1–8, show a schematic description of the installation of a ladle. The INSURAL 10 backing material comprises three components, which should be blended in dry form. A suitable level of INSURAL 10 backing is then poured into the bottom of the steel outer and the INSURAL 140 liner placed on top. Further INSURAL 10 material is then poured around the outside and tamped into place. If a pouring spout is required, a

Table 5.2   INSURAL ATL case study

Gravity die foundry with central melting and 300 kg capacity transfer ladle

Original practice: Gas heated ladle – life 12 months

	Cost £
Material: 12 × 25 kg mouldable refractory	185.00
2×25 kg wash coat	39.00
Labour for reline	156.00
Cost for gas heating (100 hrs/week)	3120.00
Total	3500.00 per year

New practice: INSURAL ATL ladle – life 5 months

INSURAL 140 liner	390.00
INSURAL 50 launder	72.00
INSURAL 80 mastic	39.00
INSURAL 10 backing	74.00
Labour for fitting	44.00
Total	619.00 for 5 months × 12/5
	= 1485.60 for 12 months
Cost for gas heating (5 hrs/week)	156.00
Total	1641.60 per year

Saving for each 300 kg ladle: £1641.60 per year

Note: The above costs are based on UK prices in 1996.

range of INSURAL 140 launders are available, one of which could be fitted by cutting a profile from the liner and sealing around the outside of the launder with INSURAL 80 mastic. When the spout is fitted and the INSURAL 10 backing is tamped down firmly, heat is applied to the steel outer using a gas torch. Once the INSURAL 10 reaches 130°C, an exothermic reaction will occur, which will hold the liner under compression. After around 30 minutes' heating, the backing will be fully cured and the top of the ladle can be finished with INSURAL 80 mastic. The ladle is now ready for use with no risk of moisture pick-up from damp refractories.

Figures 5.2a and b show a comparison between the conventional process and the INSURAL ATL lining system. It is common to find a temperature drop of 8°C per minute with traditional ladles whereas only 2.5–3°C per minute is normal using the INSURAL ATL lining system. To reduce temperature loss and avoid the possibility of hydrogen pick-up, traditional ladles are heated between transfers. INSURAL ATL ladles do not need preheating, resulting in energy conservation savings.

Table 5.2 shows a case study from a foundry which previously lined its ladles with refractory concrete and heated with gas continuously. The old style lining gave a life of 12 months.

The new practice is now to use INSURAL ATL linings, changing them every five months. The only preheating now required is that once per week any moisture absorbed over the weekend is removed.

Table 5.3   INSURAL ATL one piece ladle liners (for dimensions see Fig. 5.3)

Type	A (mm)	B (mm)	C (mm)	Nominal capacity	Euro designation
KC5294	250	185	265	10 kg	
KC5240	342	260	285	30 kg	
KC5350	293	250	363	30 kg	
KC7910	300	250	450	40 kg	
KC6294	375	260	351	40 kg	
KC5164	297	217	495	45 kg	ATL50
KC7156	470	405	462	90 kg	ATL100
KC5434	475	418	560	130 kg	ATL150
KC7804	570	540	500	170 kg	ATL200
KC7764	574	540	700	270 kg	ATL300
KC7986	730	680	725	370 kg	ATL400
KC7988	750	680	975	570 kg	ATL600
KC7990	910	840	850	770 kg	ATL800
KC7922	920	840	1000	950 kg	ATL1000

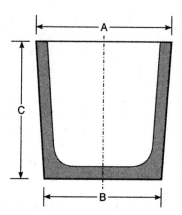

Figure 5.3   *Dimensions of INSURAL ATL one piece ladle liners (Table 5.3).*

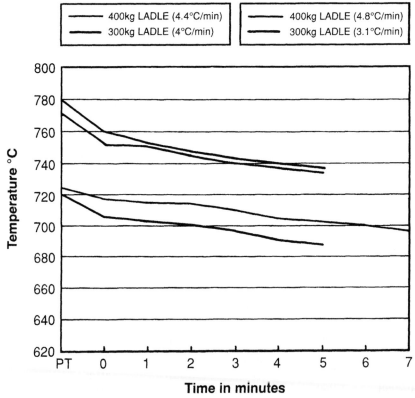

Figure 5.4   *Temperature loss with INSURAL ladle lining system.*

Benefits from using the INSURAL ATL ladle lining system for aluminium are:

Gas consumption is reduced significantly by up to 90% compared to conventional ladle practice which requires preheating
Lower melting furnace temperatures
Clean, oxide-free ladles

INSURAL ATL linings provide a clean, easily installed system, offering significant energy and cost savings. A full range of linings is available from 10 kg to 1000 kg in capacity, Table 5.3.

Figure 5.4 shows typical temperature losses with the INSURAL Ladle Lining System.

When particularly high erosion is found, such as where metal is to be poured from a great height or at a very fast rate, linings can be supplied with reinforced bases made from a FOSCAST material.

INSURAL preformed launders are used in foundries both for tapping from the melting furnaces and for the short launders feeding the holding furnaces used at each diecasting machine in a pressure diecasting foundry. The non-wetting properties of INSURAL ensure that only thin skulls of metal are left after each pour. These are easily removable, leading to low non-recoverable metal losses.

Chapter 6
Treatment of aluminium alloy melts

Introduction

Before casting aluminium alloys, the molten metal must be treated in order to:

Degas	Molten aluminium contains undesirable amounts of hydrogen which will cause porosity defects in the casting unless removed
Grain refine	Mechanical properties of the casting can be improved by controlling the grain size of the solidifying metal
Modify	The microstructure and properties of alloys can be improved by the addition of small quantities of certain "modifying" elements

There are various ways of carrying out these treatments, the older methods involve the addition of tablets or special fluxes to the melt in a ladle or crucible. In recent years, special "Metal Treatment Stations" have been developed to allow treatment to be carried out more efficiently.

Hydrogen gas pick-up in aluminium melts

Hydrogen has a high solubility in liquid aluminium which increases with melt temperature, Fig. 6.1, but the solubility in solid aluminium is very low, so that as the alloy freezes, hydrogen gas is expelled forming gas pores in the casting. The hydrogen in molten metal comes from a number of sources but mostly from water:

Water vapour in the atmosphere
Water vapour from burner fuels
Damp refractories and crucible linings
Damp fluxes
Oily or dirty scrap charges
Dirty or damp foundry tools

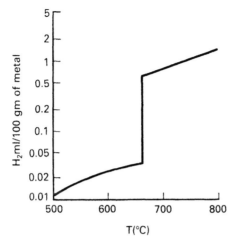

Figure 6.1 *Solubility of hydrogen in aluminium.*

To reduce hydrogen pick-up, refractories, crucibles, tools and oily scrap should be thoroughly preheated to remove water. Burner flames should be slightly oxidising to avoid excess hydrogen in the products of combustion. The melt temperature should be kept as low as possible since more hydrogen is dissolved at high temperatures. Whatever precautions are taken, however, hydrogen will still be present.

The amount of porosity that can be tolerated in a casting is determined by the method of casting and the end use of the component. If the metal cools relatively slowly, as in a sand mould, the ejected gas can build up into small bubbles which are trapped in the pasty metal. These are then uncovered by any subsequent machining or polishing operation and show as a "pinhole" porosity defect in the finished surface. The mechanical strength and pressure tightness can also be seriously affected.

Where the rate of solidification is more rapid as in gravity and low pressure diecasting, the emerging bubbles are usually small and well dispersed. They therefore affect mechanical properties less, and indeed often have a beneficial effect in offsetting possible localised shrinkage unsoundness that might otherwise cause the casting to be scrapped. For high integrity gravity and low pressure castings, it may still be necessary to apply a full degassing process.

In the past, it has not been usual to degas metal for pressure diecasting since diecastings usually contain gas porosity arising from air entrapped in the casting during metal injection. The additional porosity from hydrogen in the melt was not considered serious, particularly since the metal holding temperature for pressure diecasting is usually low, reducing the amount of hydrogen pick-up. Recently, however with the improvement in diecasting technology, more diecasters are using degassed metal (see Chapter 9).

Degassing aluminium alloys

The maximum concentration of dissolved hydrogen possible in aluminium alloys can be as high as 0.6 ml H_2/100 g. By careful attention to melting practice this can be reduced but even with the best practice, remelted foundry alloys may be expected to contain 0.2–0.3 ml H_2/100 g Al.

The degassing process involves bubbling dry, inert gases through the melt to reduce the hydrogen level to around 0.1 ml/100 g. The liquid and solid solubilities of hydrogen are different in different alloy systems and a hydrogen level of 0.12 ml/100 g will give castings free from porosity in LM4 (Al–Si5Cu3Mn0.5) while the low silicon Al–Cu–Ni alloy BSL119 will be porosity free at 0.32 ml H_2/100 g. If levels of hydrogen are taken too low, it is difficult to avoid some shrinkage porosity in the castings.

For many years, the use of chlorine gas, developed by plunging hexachloroethane in the form of DEGASER tablets, was the standard method of treatment. With effect from 1 April 1998 the use of hexachloro-ethane in the manufacturing or processing of non-ferrous metals was prohibited in European countries other than:

> For research and development or analysis purposes
> In non-integrated aluminium foundries producing specialised castings for applications requiring high quality and high safety standards and where consumption is less than 1.5 kg of hexachloroethane per day on average
> For grain refining in the production of the magnesium alloys AZ81, AZ91 and AZ92

The regulations implement Paris Commission Directive 97/16/EC. The exceptions will be reviewed later.

Because of this ruling, Foseco has withdrawn from sale all products containing hexachloroethane. Tablet degassing has been replaced by degassing with dry nitrogen or argon using a lance or preferably a specially designed rotary impeller which ensures even dispersion of fine bubbles throughout the melt. DEGASER 700 tablets which, when plunged under the metal surface produce nitrogen gas, are available in some countries and can also be used.

Rotary degassing

For any degassing technique to be efficient, it is necessary that very fine bubbles of a dry, inert gas are generated at the base of the melt and allowed to rise through all areas of the molten aluminium. The metal temperature should be as low as possible during this operation. The Foseco MDU (Mobile Degassing Unit) achieves this by introducing an inert gas into the metal through a spinning graphite rotor, Fig. 6.2. A correctly designed rotor produces a large number of small bubbles into which, as they rise through

Figure 6.2 *The Foseco Mobile Degassing Unit.*

the melt, dissolved hydrogen diffuses to be ejected into the atmosphere when the bubble reaches the surface. The rising bubbles also collect inclusions and carry them to the top of the melt where they can be skimmed off, Fig. 6.3. The graphite rotor is designed to produce the optimum bubble cloud throughout the whole of the melt (Fig. 6.4).

The unit is designed to be pushed to the area of use and can be used in holding furnaces or ladles containing between 50 kg and 250 kg of aluminium.

The MDU is brought to the furnace and the arm swung over the melt. The gas flow (usually nitrogen) is then automatically switched on, the rotor is lowered into the melt and rotation commences. The treatment time, gas flow and speed of rotation are preset for a given furnace capacity and treatment should be complete in 3 to 5 minutes.

Rotor rotation speed is around 400–500 rpm and at this speed the optimum quantity of purging gas is dispersed giving very fine bubbles, resulting in high degassing efficiency and thorough cleansing of the melt through oxide flotation. After treatment the rotor is raised from the furnace or ladle, the metal skimmed clean and is ready for casting.

Typical results of rotary degassing are shown in Table 6.1.

Foseco also supplies a larger rotary degassing unit, the FDU (Foundary Degassing Unit), which is designed as a stationary unit for foundries where central melting and treatment is used before metal is transferred to the casting station. Melts of 400–1000 kg of aluminium can be treated in times of 1.5 to 5 minutes with gas flow between 8 and 20 litres/minute. The graphite rotor has a life of 100–150 treatments according to the temperature of the melt.

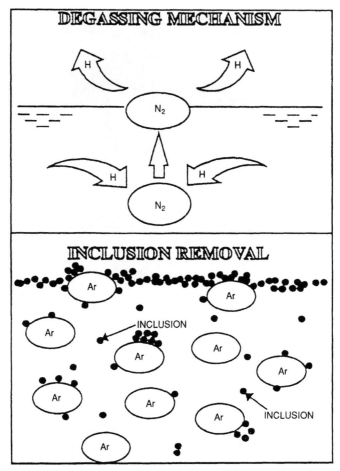

Figure 6.3   *How rotary degassing works.*

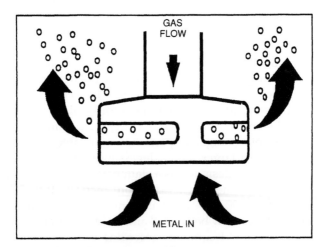

Figure 6.4   *Bubble formation in the graphite rotor*

Table 6.1 Rotary degassing of aluminium alloys

Alloy	Ladle size (kg)	Treatment time (min.)	Hydrogen content (ml/100 g)	
			Before	After
LM13 (Al–Si12Cu1Mg1)	250	5	0.27	0.18
LM25 (Al–Si7Mg0.5)	250	6	0.40	0.13
LM28 (Al–Si8Cu1.5 Mg1Ni1)	250	10	0.28	0.18
L99 (Al–Si7 Mg0.3)	220	7	0.25	0.12
L155 (Al–Cu4Si1)	90	15	0.24	0.12

The metal treatment station

A logical development of the rotary degassing system is the injection of fluxes into the melt along with the inert purge gas. Early attempts to do this were plagued with difficulty because the fluxes melted in the injector nozzles causing total or partial blockage. The introduction of granular fluxes has greatly assisted and Foseco has developed a Metal Treatment Station in which a granular COVERAL flux is introduced into a specially designed degassing rotor, Fig. 6.5. The flux feeder gives accurate dosing rates and the additive is fed into the molten aluminium at the base of the melt so that full reaction can take place before the additive reaches the metal surface.

The rotor and shaft of the Metal Treatment Station have been designed to allow the free passage of the additive into the metal melt, reducing to a minimum the problem of fusion of the treatment product in the shaft. Flux is introduced into the melt during the first part of the treatment cycle followed by a degassing cycle. The combined effect of flux injection and degassing produces cleaner alloy (fewer inclusions) than degassing alone and mechanical properties, particularly elongation values, are improved. In addition, the percentage of metallic aluminium in the dross skimmed from the melt is reduced by 20–40%.

The Rotary Degassing Unit and the Metal Treatment Station are widely used in gravity, low pressure and high pressure diecasting foundries (see Chapter 9 and 10).

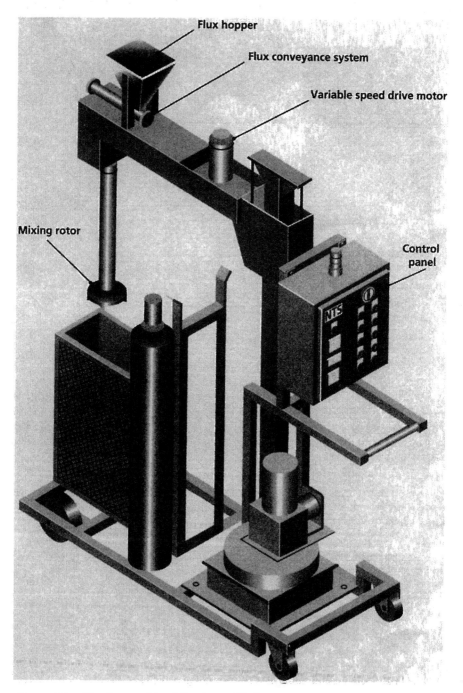

Figure 6.5 *The Foseco Metal Treatment Station.*

Grain refinement of aluminium alloys

Grain refining improves hot tear resistance, reduces the harmful effects of gas porosity (giving pressure-tight castings) and redistributes shrinkage porosity in aluminium alloys. The grain size of a cast alloy is dependent on the number of nuclei present in the liquid metal as it begins to solidify and on the rate of undercooling. A faster cooling rate generally promotes a smaller grain size.

Additions of certain elements to aluminium alloy melts can provide nuclei for grain growth. Titanium, particularly in association with boron, has a powerful nucleating effect and is the most commonly used grain refiner. Titanium alone, added at the rate of 0.02–0.15% as a master alloy, can be used but the effect fades within 40 minutes. The addition of boron together with titanium produces finer grains and reduces fade.

Titanium and boron additions may be added as a master alloy or as a flux. In the wrought aluminium industry, the benefits of TiBAl master alloys are well known, with alloy rod used in continuous applications. The majority of foundries use smaller melting and holding furnaces and continuous application is not possible so batch applications are used. While the master alloy approach has benefits of precision and controllability, salt flux addition methods are still widely used because of their convenience and low cost.

The level of silicon in the alloy affects the grain-refining response to Ti and B. Higher silicon casting alloys require higher additions of grain refiner. Typical addition levels are shown in Table 6.2.

Table 6.2   Ti additions (as TiBAl 5:1) for grain refinement of Al–Si alloys. (From Spooner S.J., Cook R., *Foundryman*, **90**, May 1997, p. 169)

Si content of alloy (%)	Ti addition (%)
4–7	0.05–0.03
8–10	0.03–0.02
11–13	0.02–0.01

The NUCLEANT range of grain-refining flux tablets is shown in Table 6.3. Most of the salts used in these products are slightly hygroscopic and an exposed tablet may pick up some moisture from the atmosphere which could increase the hydrogen content of the alloy. It is usual therefore to degas during or after nucleation.

Using the Mobile Degassing Unit, NUCLEANT self-sinking tablets are placed in the liquid metal and the rotor lowered. Grain refining and degassing take place simultaneously.

Table 6.3 Grades of NUCLEANT flux tablets

NUCLEANT	Form	Alloys treated	Application rate (%)	Remarks
2000	50 g tablets	All Al casting alloys including those with Mg	0.15–0.25	Strong grain refinement from Ti:B 6:1
70	30 g tablets	All Al casting alloys including Al–Mg alloys	0.06–0.08	Low fume similar to NUCLEANT 2000
70SS	75 g tablets	ditto	ditto	Self-sinking tablets
TILITE 101	500 g tablets	Bulk-melted Al wrought and casting alloys	0.0125–0.015	Large self-sinking tablets for the bulk melter Ti:B 10:1 one tablet can refine up to 200 kg of alloy

Refinement of hypereutectic alloys

Al–Si alloys containing over 12% Si are used for their wear resistance and it is important for consistent casting properties that primary silicon is evenly dispersed throughout the casting. With long solidification ranges, growth and flotation of primary silicon particles may occur. Large silicon particles are detrimental to castability, machinability and mechanical properties. Refinement of the structure is therefore desirable.

Hypereutectic alloys are refined with phosphorus additions of 0.003–0.015%. The aluminium phosphide formed provides nucleation sites for primary silicon ensuring a fine dispersal of silicon in the eutectic matrix. Phosphorus was conveniently added in the form of NUCLEANT 120, but this product contains hexachloroethane and has been withdrawn. New products are being developed. It is possible to use an aluminium–phosphorus master alloy for refinement.

Grain refinement by use of master alloys

A master alloy containing titanium and boron in the ratio 5:1 has the optimum effect. Master alloys are supplied in the form of rods chopped into lengths weighing 200 g which dissolve quickly and completely in the melt. Suitable additions of rod are added to the melt when the ladle is in position at the Mobile Degassing Unit. The rod melts quickly, usually by the time the rotor is inserted into the melt. Grain refinement and degassing take place simultaneously. Any increase in hydrogen content caused by the addition of the master alloy will be removed at once by the rotary degassing. The normal degassing temperature is used. The addition rate is made according to Table 6.2.

Modification of aluminium alloys

The composition of the alloy and the choice of casting process affect the microstructure of the aluminium alloy castings. The microstructure can also be changed by the addition of certain elements to aluminium–silicon alloys which improve castability, mechanical properties and machinability.

Sand cast and gravity diecast (permanent mould) alloys cool relatively slowly, resulting in a coarse lamellar eutectic plate structure which is detrimental to the strength of the castings. Changing the chemical composition to alter the microstructure is called "modification". The addition of sodium or strontium modifies the cast microstructures to give finely dispersed eutectic fibres and the coarse crystalline fracture of the alloy is refined to a fine, silky texture. These changes are accompanied by a considerable improvement in the mechanical properties of the alloy. The effect of modification on the microstructures of aluminium alloys is shown

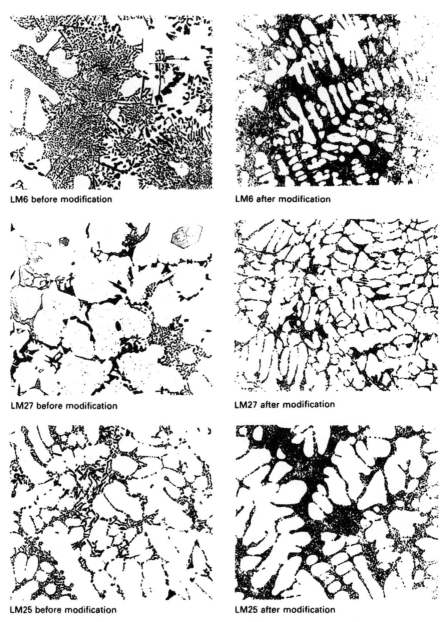

LM6 before modification   LM6 after modification

LM27 before modification   LM27 after modification

LM25 before modification   LM25 after modification

Figure 6.6   *The effect of modification on the microstructures of aluminium alloys (× 125).*

in Fig. 6.6. Modification increases hot tear resistance and alloy feeding characteristics, decreasing shrinkage porosity.

Pressure diecastings are rapidly cooled in the mould giving small grain size with a fine eutectic structure with small dendrites. Modification of pressure-diecast microstructures is also possible and the lamellar eutectic silicon will be changed to a fine fibre structure.

The higher the silicon level in an alloy, the more modifying element is needed to change the structure. The faster the freezing rate, the lower the amount of modifier required. The first hypoeutectic modifiers were based on sodium and they are still widely used today although "fade", the gradual loss of sodium with time, can lead to problems of control. Sodium has a very large undercooling effect so that it is particularly useful in slowly cooled casting processes such as sand casting. Because of its reactivity, sodium is vacuum packed in aluminium containers for convenient addition. Sodium-based fluxes may also be used.

Strontium as a modifier has the advantage over sodium that it is less reactive and can be added in the form of master alloys so that precise control over additions is possible and fade only occurs over a period of several hours but it is less effective in heavy section castings.

Hypereutectic alloys must be modified with phosphorus, resulting in a fine primary silicon particle size.

Sodium modification

Sodium can be added either as metallic sodium in the form of NAVAC or as sodium salts in the form of COVERAL flux 29A or 36A or GR2715 granular flux. The salts process, although slower in terms of sodium transfer, is less likely to introduce gas into the melt. NAVAC is a pure form of metallic sodium, vacuum sealed, free from oil- and other gas-producing impurities and introduces a minimum amount of gas.

The modified structure is unstable and tends to fade, that is, to revert to the unmodified condition. The rate of reversion depends on silicon content, temperature and size of the melt. Reversion is slow at temperatures below 750°C and does not occur to any considerable extent during a 10 minute holding period as long as the metal is not agitated by stirring or degassing.

Salts method

When degassing with DEGASER tablets, it was always considered advisable to degas before sodium modification, since degassing with hexachloroethane removes sodium from the melt. The Rotary Degasser removes much less sodium, but modification using flux is still best carried out after degassing.

The charge should be melted under COVERAL 11 or GR2516 and heated to above 750°C then degassed with the Mobile Degassing Unit for 3–5 minutes. The melt is skimmed and COVERAL 29A (for temperatures of 790–800°C), COVERAL 36A (for temperatures of 740–750°C) or granular flux GR2715 added and worked into the melt for 3 minutes. Leave for 5–10 minutes to allow the melt to reach the required pouring temperature, dross-off with COVERAL 11 or GR2516 and skim before pouring.

Sodium metal method

NAVAC is a vacuum-processed metallic sodium, sealed in air-tight containers which minimises the possibility of gas pick-up. It also avoids the likelihood of undesirable crucible attack which may result from the use of modifying sodium salts. NAVAC is available in the following sizes:

NAVAC 12½, 25, 50, 100, 500

The number represents the weight of sodium in grams.

Since some possibility of hydrogen pick-up is possible, it is preferable to degas after modification. The charge is melted under a layer of COVERAL GR2516 or 11. The dross is pulled to one side and when the melt is around 750°C, the NAVAC container is plunged using a bell-shaped plunger which has been coated with refractory dressing (HOLCOTE 110 is suitable).

One NAVAC 25 unit per 50 kg of melt will introduce approximately 0.05% sodium giving a residual sodium in the melt of 0.01–0.012% Na which is suitable for 10–13% Si alloys. Lower silicon alloys require less, about 25 g/100 kg of metal.

When the reaction has subsided, the metal is stirred using the plunger, but without breaking the metal surface. The plunger is withdrawn and the Mobile Degassing Unit rotor immersed, degassing takes 3–5 minutes. The melt is then drossed off with a small amount of COVERAL GR2516 or 11 and poured without delay. The modified structure fades with time, and melts which are held for longer than 10 minutes should be partially or wholly remodified from time to time with further additions of sodium.

Safety precautions

Sodium will burn fiercely in air and will react explosively with water. NAVAC should be kept in a safe fireproof store. Perforated or damaged containers should never be used.

Strontium modification

The ability of strontium to modify the structure of aluminium–silicon alloys without the effect of fading on standing has made this method of modification popular for low pressure and gravity diecastings where it may be necessary to hold molten metal for relatively long periods. Sodium can be used for this purpose but topping up additions are required every 30–40 minutes to maintain the modified structure.

Strontium is added as a master alloy containing 10% Sr for use principally on hypoeutectic and eutectic Al–Si alloys (6–8% and 10–13% silicon) and is used mainly on alloys for gravity and low pressure diecasting. Strontium

Table 6.4   SrAl additions for eutectic modification (from Spooner S.J., Cook R., *Foundryman*, **90**, May 1997, p. 170)

Si content of alloy (%)	Sr addition (%)	10SrAl Addition (kg/tonne)	200 gm piglets per 100 kg melt
4–7	0.01–0.02	1–2	0.5–1
8–10	0.03–0.04	3–4	1.5–2
11–13	0.04–0.06	4–6	2–3

10SrAl modifier is supplied in the form of "piglets" weighing 200 g. They are added to the melt as the Mobile Degassing Unit is swung into position, allowing degassing and modification to take place at the same time, Table 6.4.

Addition of one 200 g piglet to 100 kg of metal adds 0.02% Sr and, in good conditions, almost 100% yield is possible. Solution is usually complete within 3 minutes of plunging the piglets, which is within the treatment time when using the rotary degassing unit. The rate of loss of strontium from molten metal is slow and allows holding times of a few hours.

Foundry returns from strontium-modified metal contain an uncertain amount of active strontium modifier, so many foundrymen prefer to use sodium modification, knowing that no active modifier will be present in the returns.

Permanent modification

Antimony (Sb) has a permanent modifying effect on the structure of the eutectic phase. However, it is not usual to use Sb in foundry applications. There is a danger of formation of toxic stibnine gas (SbH_4) and there is a danger of overmodification when scrap is recycled. Sb addition can seriously impair the performance of Na and Sr additions.

Melting procedures for commonly used aluminium alloys

Sand, gravity die and low pressure diecasting

The popular casting alloys fall into two groups:

Medium silicon   LM4 (Al–Si5Cu, US 319)
                 LM25 (Al–Si7Mg, US A356)
                 LM27 (Al–Si7Cu2Mn0.5)
Eutectic silicon LM6 (Al–Si12, US 413)
                 LM20 (Al–Si12Cu, US A413)

Medium silicon alloys, 4–7% Si

The alloys should be melted under a covering/drossing flux and degassed. Grain refinement is advantageous and can assist response to heat treatment. Modification is not essential.

Crucible melting (using the MDU)

1 Melt under COVERAL GR2516 using 0.125–0.25% or COVERAL 11, using 0.5%, raising the temperature to 750–760°C.
2 Grain refinement can be done by plunging NUCLEANT 2000 tablets (0.25%) or NUCLEANT 70 tablets (0.08%).
3 Skim off the flux. Swing the Mobile Degassing Unit into position and degas for 3–5 minutes. As an alternative to the use of NUCLEANT, grain refinement can be carried out simultaneously with degassing by using TiB 5/1 grain refiner. Suitable lengths of rod to give 6–10 kg/tonne addition are added to the melt when the ladle is in position at the Rotary Degassing Unit. The rod melts quickly, usually by the time the rotor is inserted into the melt. Grain refinement and degassing take place simultaneously. Skim the metal clean before casting.

Bulk melting

1 Melt under COVERAL GR2220 granular flux using about 0.5 kg/square metre of melt area, or COVERAL 5F powder flux at 1 kg/square metre forming a complete cover adding half early and the rest when the charge is molten.
2 Transfer the required amount of metal to the transfer ladle, grain refine by plunging NUCLEANT 2000 tablets (0.25%).
3 Degas using the MDU. As an alternative to the use of NUCLEANT, grain refinement can be carried out simultaneously with degassing by using TiB 5/1 rod, as described above.

Suggested pouring temperatures for sand castings:

Light castings, under 15 mm	730°C
Medium castings, 15–40 mm	710°C
Heavy castings, over 40 mm	690°C

Eutectic silicon alloys, 12% Si

The alloys should be melted under a covering/drossing flux. Grain refinement benefits heavy section castings and the eutectic alloys benefit greatly from modification.

Crucible melting

1 Melt under COVERAL GR2516 or COVERAL 11 as in 1 above, taking the temperature to 750°C.
2 Modify the alloy by drawing the dross to one side and plunging a NAVAC unit (one NAVAC 25 per 50 kg of metal). When the reaction has subsided, raise and lower the plunger a few times to stir the metal gently, allow the metal to stand for a few minutes, then skim off the dross.
3 Swing the Mobile Degassing Unit into position and degas for 3–5 minutes.
4 If grain refinement is required, NUCLEANT 2000 tablets (0.25%) may be plunged before degassing. Alternatively TiB 5/1 rod can be added before degassing.
5 Skim clean before casting.

Bulk melting

1 Melt under COVERAL GR2220 granular flux using about 0.5 kg/square metre of melt area, or COVERAL 5F powder flux at 1 kg/square metre forming a complete cover adding half early and the rest when the charge is molten.
2 Transfer the required amount of metal to the transfer ladle.
3 Modify the alloy by plunging NAVAC (one NAVAC 25 unit for 50 kg of metal). If strontium modification is preferred, which may be the case if the metal is to be transferred to a holding furnace, 10SrAl master alloy can be plunged adding one 200 g piglet to 50 kg of metal (0.04%).
4 Swing the Mobile Degassing Unit into position and degas for 3–5 minutes.
5 If grain refinement is needed as well, NUCLEANT 2000 tablets (125 g/50 kg metal) can be plunged before degassing or TiB 5/1 rod added before degassing.
6 Skim the metal clean before use.

Suggested pouring temperatures for sand castings:

Light castings, under 15 mm	730°C
Medium castings, 15–40 mm	710°C
Heavy castings, over 40 mm	690°C

Treatment of hypereutectic Al–Si alloys (over 16% Si)

These are wear-resistant alloys used for pistons and unlined cylinder blocks; they may be sand, chill or pressure cast. Grain refinement is necessary to improve castability and machinability. Hypereutectic alloys must be refined with phosphorus using Al–P master alloy. Alternatively prerefined ingot can be used. Melting must be under a sodium-free flux, since sodium prevents the refining action of phosphorus. Degassing is necessary but modification with sodium or strontium is not used.

Melting practice

1 Melt under COVERAL 66 sodium-free flux, adding 0.5% with the charge and a further 0.5% when molten.
2 Bring the melt to 780°C and plunge Al–P master alloy, dross-off.
3 Degas with the Mobile Degassing Unit.
4 Skim the metal clean before use.

The casting temperature for these alloys is high, around 750–760°C.

Melting and treatment of aluminium–magnesium alloys (4–10% Mg)

These alloys oxidise rapidly during melting and also pick up hydrogen readily. Traces of sodium are harmful and sodium-free fluxes should be used. Grain refinement is necessary.

Crucible melting

1 Heat the crucible and charge the solid metal dusted with a generous quantity of COVERAL 65 (250 g for 50 kg of metal) and melt rapidly.
2 At a temperature of around 600°C, when the metal is pasty, add a further quantity of COVERAL 65 (1 kg for 50 kg of metal).
3 Do not exceed 750°C, stir the fluid flux into the melt using a skimmer or plunger to contact the flux as much as possible with the metal, keep stirring until the flux has turned dry and powdery.
4 Degas using the MDU and grain refine using TiB 5/1 rod or by plunging NUCLEANT 2000 before degassing.
5 Dross-off by sprinkling COVERAL 66 (250 g per 50 kg of metal) and leaving for 5 minutes.
6 Rabble into the surface to start the exotherm and skim off the powdery dross.

Casting temperature is typically 700°C. A metal-mould reaction may occur in green sand moulds and an inhibitor such as boric acid may be incorporated in the sand. With chemically bonded sand moulds and cores, a suitable coating such as MOLDCOTE 41 may be used instead of an inhibitor.

Special requirements for gravity diecasting

Thin section (<5 mm) gravity diecastings cool so quickly that it is sometimes considered unnecessary to grain refine or modify the alloy. For castings having sections above 5 mm, grain refinement is beneficial and modification of eutectic alloys (10–13% Si) may be used. Melting practice for high quality gravity diecastings and low pressure diecastings is generally the same as for sand casting and is described above.

In some less-critical gravity diecastings, a small amount of gas in the metal may be beneficial since dispersed gas porosity may be considered less harmful than shrinkage. The use of DYCASTAL, which induces limited and controlled gassing of the melt, is frequently found useful in the prevention of shrinkage in many gravity diecastings.

The metal is melted as usual using the appropriate COVERAL flux. Degassing is not usually necessary. Immediately prior to baling out metal for casting, plunge 0.25–0.5% of DYCASTAL 40 tablets (depending on the existing gas content of the alloy and the severity of the defects to be overcome) and hold until the reaction is complete. Skim off the dross and cast. The treatment must be repeated at 20–30 minute intervals as the gassing effect lasts for only this time. If draws or shrinkage are seen on the castings being made, this indicates that further treatment with DYCASTAL 40 is necessary.

For hand ladle treatment, DYCASTAL 1 powder may be sprinkled by means of a shaker into the bottom of the preheated ladle where it melts. The ladle is filled with metal and when the bubbling ceases, the ladle should be poured without delay. The amount used must be found by experience for the particular melting practice and castings being made. DYCASTAL should not be used where the rate of solidification is slow or for high integrity castings.

Treatment of alloys for pressure diecasting

Pressure diecasting is a fast, repetitive casting process in which molten metal is injected at high pressure (e.g. 20 MPa, 3000 psi) into a steel die. High production rates are possible with wall thicknesses as low as 1 mm (though 2–3 mm is more usual). Surface finish and dimensional tolerances are excellent. The die may incorporate retractable steel cores but sand cores cannot be used because the high metal pressure would cause penetration.

With conventional pressure diecasting, castings usually suffer from porosity because air in the die cavity becomes trapped in the casting; moreover thick sections are difficult to cast sound because of shrinkage. The castings cannot be fully heat treated since the trapped air causes "blistering" of the casting surface when the casting is heated. On the other hand, the rapid chilling in the die produces a fine grain structure and good as-cast strength.

The commonly used alloys contain 8–10% Si and 2–3% Cu with around 1% Fe. The alloys are usually bulk melted and transferred to holding furnaces at each casting machine from which metal is dispensed into the die by an automatic ladle, or in manual operations, by baling out. The holding furnaces are topped up from time to time from the melting unit.

In the past, it was not considered worthwhile to degas the alloy before casting (because some gas porosity was accepted as inevitable). Furthermore, the cooling rate in the die is so fast that grain refinement and modification were also not considered necessary. However, with improvements in the practice of pressure diecasting (see Chapter 9), it is found that treatment of the liquid alloy by Rotary Degassing is effective both in degassing and removing non-metallic inclusions, so some diecasters degas in the transfer ladle for this reason. Strontium modification is also found to be beneficial.

Bulk melting is carried out under a covering/drossing flux to ensure minimum metal loss. COVERAL GR2220 granular flux at 0.5 kg/square metre is used or COVERAL 5 powder flux at 0.5–1% of metal weight or 1 kg/square metre of melt area forming a complete cover, adding half early and the rest at final melt down.

In the holding furnace, COVERAL 75 low temperature flux can be used as a cover to reduce metal loss. Scatter 0.25–0.5% onto the metal surface and rabble gently until the exotherm develops. Push aside or remove before taking ladles.

See Chapter 9 for further details.

Chapter 7

Running, gating and feeding aluminium castings

Introduction

The primary function of a gating system is to introduce clean, dross-free metal from the pouring ladle to the mould cavity and to do so in a manner which will not cause subsequent reoxidation and gas pick-up. Aluminium alloys are all subject to dross formation, a film of oxide forms immediately on any metal surface exposed to air (see Table 3.1) Oxide films form on the surface of the metal stream as it pours from ladle to mould and will form within the mould cavity as the mould is filling with metal. Turbulence of the metal stream, outside or within the mould, not only exposes more metal surface to oxidation, but also entraps oxide films within the casting (oxide folds) leading to a reduction in mechanical properties.

There is no general agreement among foundrymen about the method of gating aluminium castings although certain principles are widely accepted:

Where possible, gating should be into the bottom of the casting
Unpressurised gating should always be used, that is, the gate areas should not limit the flow rate into the mould cavity
Ingates should be taken from the top of the runner to ensure that the runner bar is always full
The sprue should control the fill rate of the casting
The sprue should be designed to avoid entraining air and dross, it should be tapered downwards so that the sprue base is the flow controlling area
Low stream velocities should be used to avoid turbulence, optimum stream velocities as low as 500 mm/s have been reported for Al alloys (and 75 mm/s for Al bronze)

The introduction of ceramic foam filters has, however, changed traditional ideas of gating aluminium castings.

Gating without filters

The following traditional gating rules apply to all dross-forming non-ferrous alloys including aluminium alloys and bronzes such as aluminium bronze. They are mainly applicable to sand moulds.

Pouring bush

The use of a properly designed pouring bush is recommended on all but the smallest of castings. The pouring bush should be designed in such a way so that the pourer can fill the sprue quickly and maintain a near constant head throughout the pour. An offset design incorporating a weir achieves this objective, Fig. 7.1. The pouring bush should be rectangular in shape so that the upward circulation during pouring will assist in dross removal. The exit from the pouring bush should be radiused and match up with the sprue entrance.

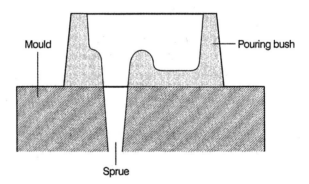

Figure 7.1 *A properly designed pouring bush.*

The practice of pouring directly down the sprue or the use of conical-shaped bushes which direct flow straight down the sprue is discouraged as not only will air and dross be entrained and carried down into the system, but also the high velocity of the metal stream will result in excessive turbulence in the gating system.

Sprue

The sprue controls the fill rate of the casting and hence is the single most important part of the gating system. Whenever production practices permit, the sprue should be tapered with the smaller controlling area at its base, all subsequent parts of the gating systems are determined from the sprue exit area. The sprue should provide a 5° taper from the controlling area. In

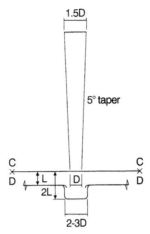

Figure 7.2 *Recommended sprue taper and sprue base dimensions.*

practice this means that the top of the sprue should have an area roughly 17 × h% greater than the bottom, where h is the sprue height.

If the sprue height is over 300 mm it is sufficient to increase section diameter from bottom to top by 50%, Fig. 7.2.

The cross-section of the sprue can be round, square or rectangular. There is evidence to suggest that a rectangular shape is to be preferred due to a reduced tendency to vortex formation which could result in air aspiration.

Sprue base

Because stream velocity is at its maximum at the bottom of the sprue it is important that a sprue base be used to cushion the stream and allow the flow to change from vertical to horizontal with a minimum of turbulence. Recommended sizes of the sprue base are a diameter 2–3 times the sprue exit diameter and a depth equal to twice the depth of the runner bar, Fig. 7.2.

Runners and gates

Non-ferrous alloys should always be cast with an unpressurised gating system with the runner in the drag (lower half of the mould) and the ingates in the cope (upper half of the mould). The area of the runner bar should be between 2 and 4 times the area of the sprue base, and the total area of ingates at least equal and up to twice the runner area. This is to ensure that the required fill rate is achieved at the lowest possible velocity. Alloys particularly susceptible to drossing, such as aluminium bronze, may require even larger runners and gates to ensure stream velocities are kept to a minimum.

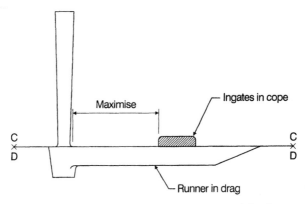

Figure 7.3   *Maximise the distance between the sprue and the first gate for effective inclusion removal.*

Runner cross-sections should ideally be rectangular, with a width to depth ratio of 2:1, the wider upper surface is to maximise the potential of the runner bar to collect dross and inclusions. The distance between sprue and the first gate should be maximised for effective inclusion removal, Fig. 7.3. When multiple gates are taken off a runner bar, the area of the runner should be reduced by the area of each gate as it is passed, to ensure that flow from each gate is uniform. It is also good practice to incorporate a sump or dross trap at the end of runners to take the initial heavily oxidised metal.

Ingates should enter the mould cavity at the lowest possible level to avoid turbulence associated with a falling metal stream. As with the runner bar, ingates should be rectangular in cross-section rather than square so as not to cause a "hot-spot" and subsequent porosity at the casting contact. The exact width to thickness ratio should be determined by the solidification time of the casting; as a rule of thumb, the thickness of ingates should be less than one-third the thickness of the casting at the point of contact, Fig. 7.4.

Rules such as these are applicable mainly to castings made in small numbers in sand moulds where there is little opportunity to experiment with running systems. The rules often conflict with other foundry requirements:

> Low stream velocities may not allow the casting to be filled quickly enough to avoid cold metal defects in thin sections of the casting
> Large area runners and gates lead to poor yields which may be economically unacceptable
> Bottom gating means that the coldest metal is at the top of the casting, just where the hottest metal is needed to ensure feeding to avoid shrinkage defects in the casting

Recent studies using real-time X-radiography, have allowed actual flow patterns occurring in running systems to be studied with some surprises for the foundryman. The filling basin, if it is a simple cone, allows metal to fall

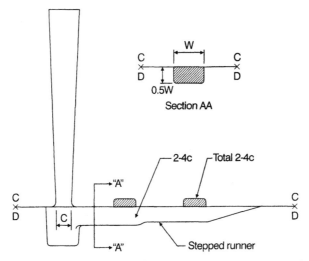

Figure 7.4 *Recommended runner and gate area, runner system and stepped runner.*

unchecked from the pouring ladle leading to fast filling of the mould and allowing air bubbles to be carried through into the mould cavity. In gravity dies, in particular, these bubbles may become trapped causing casting defects. Even if the sprue is carefully designed, with correct inlet and outlet areas and the correct taper, the initial phase of filling the sprue can trap air bubbles and carry them into the running system.

The well at the base of the sprue was thought to assist the turn of the metal through the right angle bend and control the entry of metal into the runner, but in fact even the best designed well contributes to disintegration of the flow and introduces air and oxide.

The runner, before it fills completely with metal, may also entrain air as a high speed jet of metal flows initially along the base of the runner, reflecting from the far end and rolling backwards over the underlying fast jet. Air is trapped between the opposing flows feeding bubbles into the mould cavity.

Ingates, placed on top of the runner, are expected to fill only after the runner fills. This is not so, metal can enter the gate at an early stage and spray from them at high velocity into the mould cavity.

Gating with filters

The widespread use of foam ceramic filters has introduced a new dimension into the running and gating of aluminium castings. Filters have several important effects:

They effectively trap dross and some oxide films
They control metal flow rate
They reduce turbulence

The use of ceramic filters allows the traditional gating rules to be modified while still achieving quality castings. Foam ceramic filters have a distinct advantage over the extruded type in that there is no separation of the initial metal stream which passes through them, hence the possibility of reoxidation at the filter exit face is less. The provision of a ceramic foam filter immediately after the base of the sprue changes the flow patterns dramatically.

The filter requires a certain amount of pressure and time to prime, so the flow of metal is temporarily arrested on encountering the filter, this allows the sprue to backfill excluding air from the incoming metal. Metal emerges from the exit of the filter in a single turbulence-free stream at low velocity, hence the runner fills gently and the gates operate as designed. The casting then fills without the entrainment of air and oxide films.

The beneficial effect of filters is seen therefore as mainly due to their ability to eliminate turbulence, although they also filter out any gross dross inclusions carried over from the melting unit.

Full details of the use of filters are given in Chapter 8.

Feeding mechanisms in Al alloy and other non-ferrous castings

Aluminium alloys shrink by 3.5–6.0% during solidification, so that without feeding, castings must contain porosity defects. The feeding requirements are dependent to a large extent on the freezing range of the alloy being cast, Table 7.1.

The freezing ranges of other non-ferrous alloys are given in Chapter 1.

Table 7.1   Approximate freezing range of Al alloys

Alloy		*Approximate freezing range (°C)*		
LM0	Al 99.5	657–650	7°	⎫
LM6	Al–Si12	575–565	10°	⎬ Short freezing range
LM20	Al–Si12Cu	575–565	10°	⎭
LM4	Al–Si5Cu3	580–520	60°	⎫ Long freezing range
LM25	Al–Si7Mg	615–550	65°	⎭

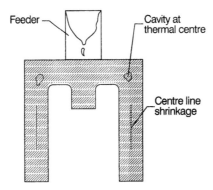

Figure 7.5 *Typical forms of porosity in short freezing range alloys.*

Solidification of short freezing range alloys (Fig. 7.5)

When an alloy of short freezing range cools in a sand mould, that portion of the liquid which first reaches the liquidus temperature begins to solidify. This usually occurs at the mould interface where heat extraction is greatest. The chilling action of the mould wall results in the formation of a thin skin of solid metal surrounding the liquid. With further extraction of heat through this shell of solid metal, the liquid begins to freeze onto it and the wall of solid metal increases in thickness. The solid and liquid portions are separated by a relatively sharp line of demarcation – the solidification front – which advances steadily towards the centre of the casting. The crystal growth on the solidification front is relatively short and corresponds to the start of freeze at their apex and the end of freeze at their bases. Short freezing range alloys encourage directional solidification even at relatively low thermal gradients.

Solidification of long freezing range alloys (Fig. 7.6)

With long freezing range alloys, the development of directional solidification is difficult. Although a thin skin may initially form on the mould walls, solidification does not proceed progressively inwards. Instead, solidification begins by the advance from the mould walls towards the interior of a "nucleation wave" corresponding with the liquidus isotherm. At some time later a second "end-of-freeze wave" corresponding with the solidus isotherm moves away from the mould walls and pursues the "nucleation wave" towards the centre of the casting. Freezing thus begins at each location in the casting when the nucleation wave passes it and ends there when the end-of-freeze wave reaches it. In general there are three distinct zones during solidification of a long freezing range alloy: a completely liquid zone at the thermal centre of the casting; a zone of solid metal next to the mould walls and a region of partial solidification between the liquid and solid zones.

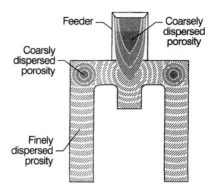

Figure 7.6 *Typical forms of porosity in long freezing range alloys.*

Factors which influence solidification mechanisms

There are a number of factors which affect the solidification mode of a particular alloy. The solidification range of an alloy, as measured in temperature degrees, is not on its own a true indicator of solidification mode. Rather the time interval between the start of freeze and the end of freeze is what determines how an alloy will solidify. The interval between liquidus and solidus is determined by a number of factors including:

The solidification range of the alloy:
This is a fundamental characteristic of a particular alloy. For a given mould material, the effect of increasing solidification range is to increase the time interval between start and end of freezing,
The thermal characteristics of the mould:
An increase in the thermal conductivity of the mould material reduces the interval between start and end of freeze,
The thermal conductivity of the solidifying alloy:
The effect of an alloy of high thermal conductivity is to reduce thermal gradients within the casting, thus increasing the interval between start and end of freeze.
Solidification temperature:
A low solidification temperature which reduces temperature gradients between mould and casting will decrease thermal gradients within the casting and increase the interval between start and end of freeze.

The effect of solidification mechanism on shrinkage distribution

Generally speaking, the short freezing range alloys show deep pipes in the feeders as feed metal is supplied to the casting throughout the entire solidification interval. Internal porosity within the casting can take the form of small open cavities which occur near the end of the solidification when the feed metal is cut off by the merging of parallel solidification fronts, this is commonly called centreline shrinkage. Another form of shrinkage exhibited by short freezing range alloys is open cavities at inadequately fed thermal centres and isolated heavy sections, Fig. 7.5.

With alloys of long freezing range, feeders often show minimal pipe as the "mushy" solidification mode will only allow liquid flow for a part of the total solidification time. Finely dispersed porosity can exist throughout the entire casting section, with coarser concentrations at parts of slower cooling such as junctions and under feeder heads, Fig. 7.6. Under normal foundry conditions, it is virtually impossible to achieve absolute soundness in extremely long freezing range alloys such as tin or phosphor bronzes.

Feeding of aluminium alloy and other non-ferrous alloy castings

For satisfactory feeding of alloys of short freezing range, feeders must be placed over thermal centres of the casting and they must solidify after that part of the casting to which they are connected. Insulating feeding aids such as KALMIN 70 are used to ensure effective feeding and to improve yield. The feeders must be of sufficient volume to compensate for the liquid and solidification shrinkage of the alloy which is influenced by the alloy composition, the degree of pouring superheat, the shape of the casting and the gas content of the alloy. Due consideration should be given to the feeding range of the alloy. The methods of calculating feeder and feeding aid requirements are described in Chapter 17.

The concept of directional solidification has little relevance with alloys of long freezing range. Generally the goal in feeding such alloys is not to eliminate porosity totally but to ensure that it is dispersed as evenly as possible throughout the casting section. It is often desirable for feeders to compensate only for superheat and a portion of solidification shrinkage so as not to extend the solidification time excessively. The shrinkage volume for which the feeders must compensate is again influenced by the alloy constitution, the degree of pouring superheat, the section thickness of the casting and the gas content of the alloy. Long freezing range alloys have virtually no feeding range and under normal foundry conditions, achieving a high degree of soundness is virtually impossible.

Because of the difficulties of calculating precise feeding requirements for aluminium alloys, foundrymen still rely to a large extent on experience to achieve acceptable results.

In some copper-based alloys, such as gunmetals and phosphor bronzes with very long freezing ranges, the difficulty of feeding porosity is aggravated by high thermal conductivity, specific heat capacity and latent heat of solidification. In these cases it may be advisable to cast without feeders to keep thermal gradients as uniform as possible to encourage evenly dispersed porosity.

Simulation modelling

A number of software packages are now available which model the flow of metals into dies or moulds and allow the solidification of the casting to be simulated. Computer modelling is being increasingly used for the design of dies and moulds for aluminium casting in order to reduce the lead time required for making new castings.

Predictive fluid flow software, Magmasoft being one of the best known, uses physics-based modelling to allow mould filling to be studied and its effects on casting soundness to be assessed. Ideally such modelling should enable the onset of turbulence during mould filling to be predicted and the effect of gating systems on the temperature distribution within the casting to be studied. While flow modelling is not yet perfect, it does enable possible danger areas in the casting to be predicted.

Figure 7.7 shows a gravity diecast air intake manifold casting. The original gating system (Fig. 7.7a) had 4 ingates spaced evenly along the length of the casting. The casting had severe pressure tightness problems. The casting was subjected to mould filling simulation to identify and eliminate the defect. The first step in any flow modelling investigation is to obtain a 3D CAD model of the mould cavity and all of the boundary conditions such as alloy type, mould and core materials, coatings used etc. The filling simulation indicates the direction of flow of metal at any point with arrows. The velocity of the metal is indicated by the arrow length and the arrow colour shows the temperature for any given moment in time during the filling sequence.

Figures 7.7b–e show the progress of mould filling in the first 3 seconds.

Figures 7.8a–f show the same manifold rigged for direct pouring through a foam filter located to combat the problem. The mould filling modelling can also be used to help design pressure diecasting dies, even for components as complex as an aluminium engine block. Process parameters such as pouring temperature, shot profile of the diecasting machine piston, cycle times, die-cooling water flow and temperature are taken into account. Even with filling times as short as 120 milliseconds, temperature distributions at the end of filling can be estimated, indicating where cold laps might be expected so that the die cooling can be modified to avoid the problem. Solidification times within the casting can also be estimated, indicating where isolated hot spots occur which could give rise to porosity problems. Metal sections can then be increased locally to optimise feeding paths.

This type of analysis will be used increasingly in the future to identify and eliminate potential sources of defects such as hot spots, cold laps, misruns and oxide defects. Feed metal requirements will be quantified and optimum pouring temperature proposed. Increasingly it will enable dies and moulds to be designed "right first time".

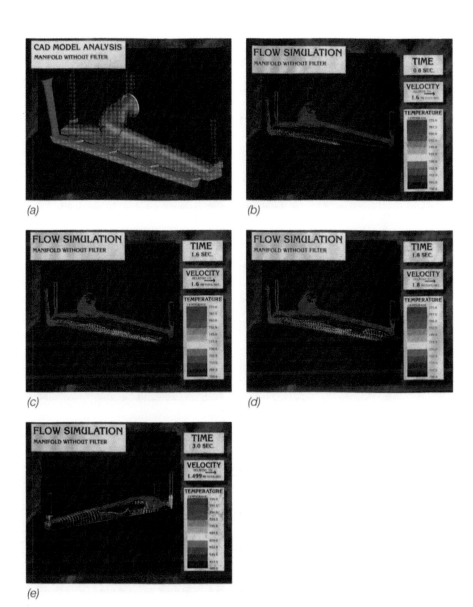

Figure 7.7 *Flow modelling of a gravity diecast air intake manifold: (a) The original gating system. (b) At 0.8 seconds, there is high velocity and turbulence in the sprue. Also hot spots and cold areas likely to become misruns can be seen. (c) At 1.6 seconds the runner bar is not yet full but metal is starting to enter the casting cavity. (d) At 1.8 seconds the runner bar is full and metal is jetting into ingate number 4. (e) At 3.0 seconds the casting is close to being full. The model has been rotated to show two metal streams not just coming together, but colliding. The two streams, being covered with oxide, do not fuse fully causing the lack of pressure tightness. In addition the risers are seen to be cold.*

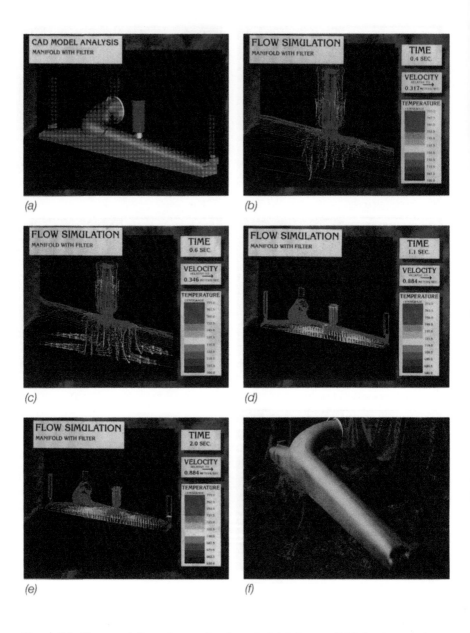

Figure 7.8 *Flow modelling of a gravity diecast air intake manifold direct poured through a filter: (a) The manifold rigged for direct pouring through a foam filter.(b) At 0.4 seconds into the pour there is low velocity metal exiting the filter. (c) At 0.6 seconds metal is flooding into the mould cavity at less than the critical 0.5m/sec. (d) At 1.1 seconds the casting is partly full with no jetting apparent.(e) At 2.0 seconds the metal is flowing smoothly outward from the central position with no opposing metal fronts. It can also be seen that there is potential for reducing the size of the risers significantly. (f) The direct poured manifold. The defect has been eliminated.*

Chapter 8

Filtration of aluminium alloy castings

Introduction

It has been known since aluminium was first used for commercial casting purposes that a melt can contain many non-metallic particles, films, or clusters in sizes from a few microns to several millimetres, Table 3.1. Whatever the size and chemical composition of these inclusions, they have been shown to be detrimental to the finished casting; decreasing the mechanical properties, increasing the propensity to leak under pressure and reducing machinability. Furthermore, these inclusions add to the difficulty of making the casting by reducing the fluidity of the metal. It has also been recognised that, since aluminium oxidises very readily, turbulence of the melt should be avoided. Turbulence leads to "folding in" of oxides and creation of new oxides from exposure of clean aluminium to the atmosphere. To counter this it is usual to cast aluminium using non-pressurised systems, for example 1:2:4 (sprue:runner:gate area) so that metal front velocities are minimised (See Chapter 7).

The introduction of ceramic foam filters to the aluminium industry in the 1970s was a major advance. The foam filter has a tortuous path through its body which traps inclusions allowing clean, smooth-flowing metal to enter the mould cavity, Fig. 8.1. By the 1980s most aerospace parts and many high integrity automotive parts were filtered.

At first, the most widely used foam filter for aluminium was made of phosphate bonded alumina ceramic, white in colour, very hard, abrasive and brittle. The brittleness caused problems due to fine particles breaking off and entering the mould cavity to form inclusions in the casting. Ceramic alumina inclusions lower the mechanical properties of the casting and can damage machine tools. Furthermore the remelting of sprues and runners containing used filters caused difficulties due to phosphorus from the phosphate bonded alumina filters contaminating the melt and possibly interfering with modification treatment. Some foundries found it necessary to cut out the portion of the runner containing the filter to be remelted separately from the bulk of the foundry returns.

Some filters were made of silicon carbide but they were also subject to some limitations in their remelt capability, since low silicon alloys can pick up silicon from remelted returns containing silicon carbide filters.

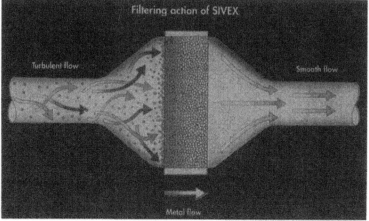

Figure 8.1   *A schematic view of the cleaning effect and flow smoothing of foam filters.*

SIVEX FC filters

To avoid these problems Foseco has developed SIVEX FC, a graphite-based non-ceramic foam filter for use with aluminium alloys. SIVEX FC filters have all the advantageous characteristics of standard reticulated foam filters, cleaning the metal by removing dross and non-metallic inclusions. Because of the filter's high surface area, even particles smaller than the size of its pores can be captured and retained in the depth of the filter. The foam structure also provides smooth, non-turbulent metal flow, so that oxide formation during mould filling is reduced. This allows simplification of gating systems, providing significant cost savings through yield improvement. Due to its method of manufacture, dimensional control is better than the older alumina filters, with typical tolerances of +0.0, −1.5 mm. Metal flow rates and flow capacity before blockage are comparable to those of similar ceramic foam filters. Specific advantages of SIVEX FC filters include:

> The absence of abrasive ceramic components allows significant new freedom in filter location, even very close to the casting surface, since the filter can be removed by machining without damaging the cutting or machining tools.
> The filter's low density – only about 60% the density of comparable ceramic filters – means that filters readily float to the surface of the melt during recycling of running systems. They are easily skimmed off with the dross, avoiding refractory particles remaining in the melt.
> The phosphate-free binder eliminates concerns about melt contamination during recycling.

The graphite composition is more easily wetted by molten aluminium than typical filter ceramics, so the metal head height required to effectively prime the filter in the early stages of pouring is reduced. With ceramic filters, a priming head of about 100 mm is typically needed. With SIVEX FC filters, this is reduced to 60 mm.

The low density and low thermal capacity of the filters result in reduced heat loss from the metal to the filter, preventing premature freezing.

Use of filters in conventional running systems

The flow of metal through a filter is shown schematically in Fig. 8.2. Initially, there is a delay while the filter is primed; no flow occurs until sufficient pressure is created by a suitable head of metal, an initial surge of metal is then observed, followed by a steady flow until filter blockage occurs. The running system must be designed to fill the mould cavity before the blockage stage is reached. The presence of the filter ensures that the lower part of the sprue and part of the runner bar are filled before metal begins to flow, thereby avoiding turbulence and air entrapment.

For most applications, the best site for the filter will be self-evident. It should be placed as near to the casting as possible in a special print so that while as much filter surface as possible is open to the metal stream, the edges are sealed and metal cannot leak round. The print can be horizontal

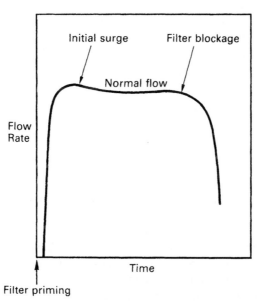

Figure 8.2 *Schematic pattern of flow through a ceramic foam filter.*

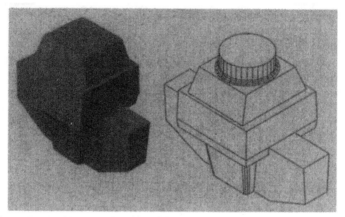

Figure 8.3 *Typical filter print for horizontal filter position.*

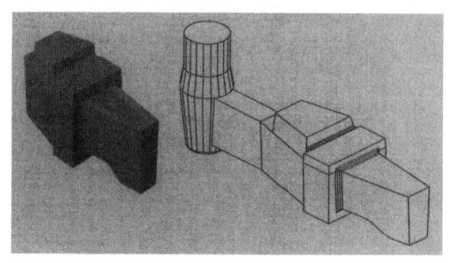

Figure 8.4 *Typical filter print for vertical filter position.*

on the joint line, with metal passing either up or down through the filter or it may be positioned in the vertical plane across a conventional runner, Figs 8.3, 8.4.

In all cases, the runner should be opened out prior to and after the filter so that almost the full area is utilised and the filter does not become a choke, Fig. 8.5. Very little pressure or metallostatic head is required to prime the filter, no special arrangements need to be made and no preheating is required. Alternative locations in the running/gating system or in the down-sprue may be preferred.

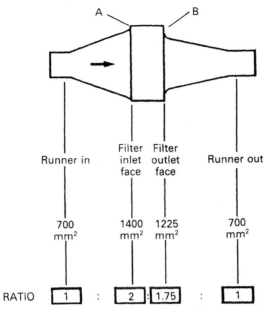

Figure 8.5 *Runner and filter dimensions for SIVEX FC ceramic foam filters for aluminium castings.*

Table 8.1 SIVEX FC filter sizes and recommended print sizes (available in 10 and 20 ppi)

Filter size (mm)	Filter area (mm²)	Recommended print on Entrance "A" (mm)	Recommended print on Exit "B" (mm)	Largest runner bar area (mm²)
30 × 30	1500	3	5	750
35 × 35	1225	3	5	613
50 × 50	2500	3	5	1250
50 × 75	3750	3	5	1875
50 × 100	5000	3	6.5	2500
75 × 75	5625	3	8	2813
25 dia.	491	3	3.5	246
30 dia.	707	3	3.5	354
40 dia.	1257	3	5	629
45 dia.	1590	3	5	795
50 dia.	1964	3	5	982
55 dia.	2376	3	5	1188

The amount of metal that can be passed through a filter depends on a number of factors, especially the type of alloy and its intrinsic cleanliness. In an existing running system, the filter area should be twice that of the existing runner bar area. In new applications the following capacity and flow rate should be considered in designing the gating system:

Capacity $1-2\,kg/cm^2$ of filter area
Flow rate $0.05-0.10\,kg/cm^2$ per second

These figures are only a guide as they depend on application, alloy type, temperature, metal head and metal cleanliness.

SIVEX FC ceramic filters are available in 10 and 20 ppi (pores per inch) offering different filtering efficiencies for varying applications. The SIVEX FC filter should be printed on all four sides as shown in Fig. 8.5.

The range of SIVEX FC filter sizes and recommended print sizes is shown in Table 8.1.

Printing to these designs ensures that no bypass of the filter will occur.

Direct pouring of aluminium alloy castings

The concept of direct pouring into the top of a mould cavity has long been recognised as desirable, with the potential benefits of:

Improved yield
Simplified sprue, gating and feeding design
Reduced fettling costs

Unfortunately, direct pouring was frequently found to introduce defects due to the turbulent flow of the metal in all but the simplest of castings. In addition, the impingement of high velocity metal streams caused erosion of moulds or cores. These objections can be overcome by pouring the metal through a ceramic foam filter situated at the base of the sprue. Clean metal, free from turbulence and oxide, fills the mould cavity and helps to feed the casting through the filter. Directional solidification and casting soundness is promoted and gates are unnecessary. The impingement problem is reduced because the metal velocity is reduced as it passes through the filter. Figure 2.1 shows the spread of tensile strength found in a series of over 100 test bars cast in Al–Si7Mg (LM25) alloy in resin bonded sand moulds. The unfiltered castings show a few, but very significant, low strength test pieces which microscopic examination of fracture surfaces showed to be due to oxide inclusions. No such low strength test pieces were seen with the filtered test pieces. Filtration has effectively removed the oxide inclusions. Remarkably, the test bars top poured through a ceramic foam filter gave consistently better mechanical properties than filtered bottom-filled castings.

KALPUR combined sleeve and SIVEX FC filter for aluminium castings

The KALPUR unit Fig. 8.6 was developed specially for aluminium sand and gravity diecastings. The direct pouring unit is positioned in a suitable location in the top of the mould or die, on the site of a feeder, for example. The location should be chosen so that the minimum free fall of metal after the filter occurs to avoid splashing. Also thought must be given to the thermal aspects of pouring so that hot spots which cause porosity are not created.

With direct pouring, the first metal in the mould or die stays at the bottom of the casting, and hot metal goes into the risers so they become more efficient. Due to the improved thermal gradient, it is possible to pour at lower temperatures than normal, which helps to reduce oxidation and

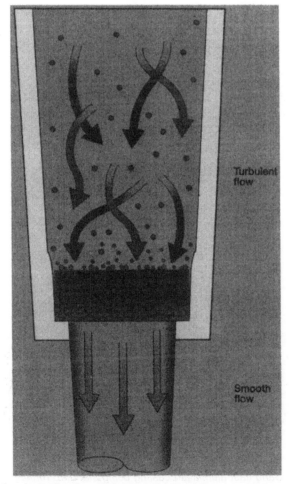

Figure 8.6 *A schematic view of the cleaning and flow-smoothing effect of pouring directly through a KALPUR unit.*

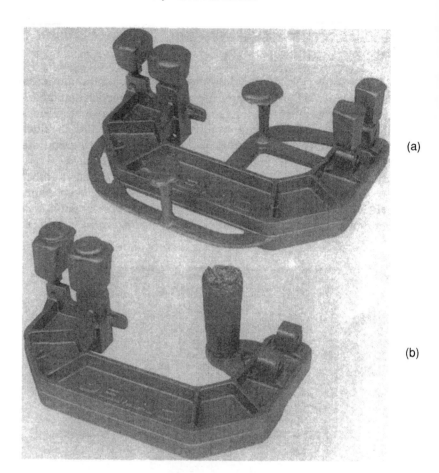

(a)

(b)

(c)

Figure 8.7   *(a) Sand cast aluminium casting using standard running system. (b) Casting made using the KALPUR direct pouring sleeve/filter unit. (c) Optimised direct pouring system.*

reduces energy costs. Gates and sprues are unnecessary which enables overall mould or die dimensions to be reduced.

An example of the value of using the KALPUR combined sleeve/filter unit is shown in Fig. 8.7. The sand casting is an element from a conveyor system weighing 3.5 kg, cast in Al-7SiMg. Figure 8.7a shows the conventional running system using 6 ingates for bottom filling. Figure 8.7b shows the use of the KALPUR unit to replace the extensive running system as well as eliminating two of the feeders. Figure 8.7c shows the optimised direct pouring system. Without the large running system, the number of castings per pattern plate was doubled and the casting yield further improved.

Direct pouring into metal dies

The KALPUR units are of greatest value for sand casting aluminium. If used in gravity diecasting, the extended solidification time promoted by the insulating sleeve can slow the production rate. Top pouring is in this case possible by incorporating a SIVEX FC filter into the downsprue. Figure 8.8 shows a 3.4 kg radiator top tank casting made in Al–Si12 alloy (LM6) by gravity diecasting. The original gating system had a conventional down-sprue, runner bar and four equally spaced ingates introducing metal along the length of the casting. The casting suffered severe pressure tightness problems which led to 100% of production needing to be impregnated. The leaking castings were caused by two metal streams colliding as the casting filled. An oxide defect formed where the streams met causing the leakage. By pouring through a ceramic foam filter in the single down-sprue, leak-free castings were obtained and the casting yield significantly improved.

Figure 8.8 *Gravity diecast casting directly cast through a SIVEX FC filter.*

Chapter 9

Pressure diecasting of aluminium alloys

Introduction

In high pressure diecasting, usually known as pressure diecasting, molten alloy is injected under pressure into a highly accurate split metal mould. It is the most widely used casting process for aluminium alloys. Production rates are fast, the process can be highly automated, and dimensional accuracy and surface finish are excellent. Thin-walled components are possible and little or no machining is needed on the cast component since holes, grooves and recesses can be finish cast.

Aluminium alloys are cast in cold-chamber diecasting machines. The die is made of tool steel usually of two cavities into which the metal is forced. The die halves are closed and locked together hydraulically to withstand the high injection pressure. Molten metal is introduced through a pouring slot into the shot tube, then a steel plunger forces the liquid metal into the die cavity under a pressure of up to 100 MPa (1000 bar), Fig. 9.1 The pressures used are given in Table 9.1.

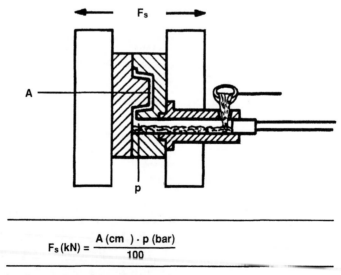

$$F_s\,(kN) = \frac{A\,(cm^2)\cdot p\,(bar)}{100}$$

Figure 9.1 *Calculation of the die opening force. (Courtesy Buhler Ltd.)*

Table 9.1 Guidance values for casting pressures (courtesy of Buhler Ltd)

| | *Casting pressure* (bars) | | |
	Al and Mg	*Zn*	*Brass*
Standard parts	up to 400	100–200	300–400
Technical parts	400–600	200–300	400–500
Pressure-tight parts	800–1000	250–400	800–1000
Chromium plating parts		220–250	

Die filling times are very short, castings with wall thickness of 3–4 mm are filled in less than 0.1 seconds. The metal solidifies rapidly because of the good thermal contact with the water-cooled die and the die set is opened to eject the finished casting together with its sprue, the process is then repeated. Cycle times depend on size and section thickness of the component, being typically 40 shots per hour for a component of 5 kg.

Machines are described by their "locking force", Fig. 9.1, which determines the cross-sectional area of the casting which can be made, which is in turn related to the overall size and weight of casting. Machines can have locking forces from 100 to over 2000 tonnes, Fig. 9.2.

Dies are expensive but can have a life of more than 100 000 shots. The process is therefore most suitable for long runs of castings. One drawback of the process is that almost inevitably some air is trapped in the die with the liquid metal, so that the casting contains gas porosity although it is frequently internal so that the component may be leak-tight. The presence of internal gas prevents subsequent heat treatment (the residual gas expands and distorts the casting) and also places a limit on the mechanical properties attained by the casting. For this reason, the process has mainly been used for castings which do not require the highest strength. Another disadvantage is that sand cores cannot be used, since the high pressure liquid metal would penetrate the core. Coring is thus limited to straight "draws" which can be formed with retractable steel "pulls".

One of the main features of recent research has been the development of special machines and processes which improve the hydraulic integrity and mechanical properties of pressure diecast components.

Die design

Dies are made of highly alloyed hot working tool steels. Because of the difficulty of machining such steels, spark erosion machining is frequently used to form the cavities. Dies must be correctly heat treated to achieve the

Control SC. Suitable for Al castings of 2.3–6.5kg shot weight. (Courtesy Buhler Ltd.)

maximum life. Die design is clearly crucial to the success of the process. Not only must the die be made to the correct dimensions, but the runner and gates must be dimensioned correctly, the die heating and cooling system must be carefully designed and the mechanical strength of the die must be designed to withstand the large forces involved. A great deal of experience is involved in the design but in recent years CAD/CAM and computer simulation of die filling and thermal conditions have been used to optimise die design. However carefully a die has been designed, it is usually necessary to try out the die, then to "fine tune" it to achieve optimum performance with regard to die temperature and production rate. Several attempts may be needed before the final design is approved and this must be allowed for in estimating the lead time needed to bring a new casting on stream.

Large diecasting foundries analyse carefully the production data from their dies to establish a relationship between casting geometry, die fill time, die temperature, casting surface quality, porosity etc. This information assists the design of new dies.

Process control

There are three stages in the process of pressure diecasting illustrated in Fig. 9.3:

1 Injection in the shot sleeve
2 Filling the die
3 Pressurisation during freezing

The internal quality of diecastings has improved in recent years through precise control of these stages. With die filling times as short as 0.1 seconds,

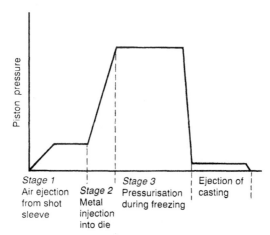

Figure 9.3 *Typical injection-control stages during pressure diecasting. (From Campbell, J. (1991) Castings. Butterworth-Heinemann, reproduced by permission of the publishers.)*

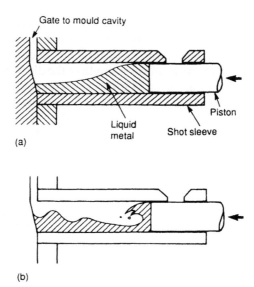

Figure 9.4 *Injection of liquid into a horizontal shot sleeve of a cold chamber diecasting machine, comparing (a) controlled and (b) uncontrolled first stages of injection. (From Campbell, J. (1991)* Castings. *Butterworth-Heinemann, reproduced by permission of the publishers.)*

metal flow speeds are very high (around 50 m/sec. gate speeds) the metal stream sprays and splashes in the cavity sealing off gas vents and trapping air within the casting.

Control of ram speed can reduce turbulence in the shot sleeve which in turn reduces air entrapment and improves casting porosity, Fig. 9.4.

Controlling the pressure profile of the injection stroke allows rapid filling of the cavity while reducing the hydraulic impact which occurs at the end of the stroke through deceleration of the plunger before the end of the die filling. This reduces flash and improves dimensional accuracy of the casting. The application of maximum pressure is delayed until freezing has started, pressure is then intensified to maintain as high a pressure as possible on the remaining liquid metal without causing the formation of flash. Transducers which measure plunger speed and pressure are used to control precisely the three stages of fill through fast, "real time", electronic control systems.

Modern diecasting machines are equipped with interactive control systems to stabilise the production conditions. Die temperature can be controlled by thermocouples set into the die which control the cooling water flow rate through the die and, where appropriate, electrical heaters. Sometimes temperature is controlled by pumping oil through channels in the die so that combined heating and cooling can be achieved. This enables the die set to be preheated before casting starts so reducing the amount of start-up scrap.

Computer monitoring of the machine parameters is frequently used. At the start of a production run, the machine controls are adjusted until plunger speed and head pressures correspond to the values which previous trials have shown to give satisfactory quality. During production runs, each shot can be monitored and all variables can be assigned tolerance limits. Warnings can be given to the operator if the process is drifting close to the preset tolerance limits. Parts made when the process is out of tolerance can be separated and identified.

The improvements in the control of the pressure diecasting process have led to significant improvements in the consistency and reliability of diecastings, which are increasingly being used for more critical applications.

Modification of the diecasting process

The pressure diecasting process is such a powerful casting process, producing castings of such excellent surface finish and dimensional accuracy, that many developments have been made to overcome the basic problem of low internal integrity of the castings due to air entrapment.

Vacuum diecasting

The shot tube and die cavity are evacuated before metal injection to reduce the amount of air that is trapped.

Pore-free diecasting

The cavity is flushed with oxygen rather than air. The oxygen should then combine with the liquid aluminium, forming oxide which is less harmful than gas entrapment.

Indirect squeeze casting

Metal is injected into a closed die cavity by a small diameter piston which also exerts sustained pressure during solidification. In the UBE system, a special thermally insulating die release agent is used which provides a significant thermal barrier between the casting and the die during filling allowing the metal to be fed further into the die cavity without freezing. When pressure is applied to the full cavity, the die coating is compressed reducing its insulating effect. The metal velocity during filling is controlled to ensure non-turbulent flow so that air entrapment is minimised. When the cavity has been completely filled, the pressure is applied. This can be from 600 to 1000 bar (60 to 100 MPa).

It is most important that the temperature gradients within the die are controlled so that the casting freezes directionally towards the gate area.

This gate area is made deliberately very thick so that it will be the last area to solidify. In this way, all the metal freezes under pressure reducing the possibility of shrinkage porosity. The low metal velocities used do not wash release agent from the die faces so that contact between the alloy and the die steel is reduced. There is no need to use high-iron alloys to prevent soldering.

The resulting physical properties of the castings are much enhanced over conventional pressure diecastings, the gas content is low so that heat treatment is possible and hydraulic integrity good. Indirect squeeze castings can be used for critical components such as brake callipers and hydraulic components. There are disadvantages: large, thin, wall castings are not possible, yield is reduced because of the large gate section that must be used, shot rates are slower than in conventional pressure diecasting and the machines are expensive.

Semi-solid metal casting

Slugs of aluminium alloy are inductively heated into a semi-solid state; the slug is introduced into the shot sleeve by a robot arm. The cavity filling process generates shear, liquefying the metal which is then injected into the die with minimum air entrapment.

All these modifications of the basic pressure diecasting process have the disadvantage of longer cycle times and more complex dies or machines, so that they are used only for special applications.

Applications of diecastings

The use of modern diecasting techniques is improving the metallurgical quality of diecastings. Heat treatment is possible with some processes so extending the application of the diecasting to components such as hydraulic manifolds, brake callipers, engine brackets, suspension arms, engine blocks etc.

One limitation of pressure diecasting is that complex cored castings cannot be made. It is not possible to use sand cores since the high injection pressure causes metal penetration. Special salt cores have been used, but are difficult to remove from the casting. Coring is limited to using tool steel "pulls".

The diecasting foundry

Diecasting foundries usually have a number of casting machines of various sizes. In addition to the diecasting machines themselves, the foundry must have melting furnaces, holding furnaces, metal treatment facilities and molten metal transport arrangements. The machines themselves need metal feeding devices to fill the shot tubes, die sprayers to coat the dies with parting agent, casting extraction devices, die heater/coolers, flash removing presses etc. All these devices can be automated to a greater or lesser extent.

Metal handling in the diecasting foundry

The commonly used pressure diecasting alloys contain 8–10% Si and 2–3% Cu with around 1% Fe. The presence of iron reduces the tendency for the casting to "solder" to the steel die but it tends to deposit inclusions in melts that are held for any length of time. The most frequently used alloys are:

Al–Si8Cu3Fe   (LM24, US A380)
Al–Si10Cu2Fe   (LM2)
Al–Si9Cu3Fe   (LM26)

Metal is normally melted in a bulk melter, either electric induction or gas-fired shaft or crucible furnaces. Liquid metal is transferred by ladle to individual holding furnaces with automatic dosing and ladling at each diecasting machine.

For many years, it was not considered worthwhile to degas the alloy before casting (because some gas porosity is accepted as inevitable). Furthermore, the cooling rate in the die is so fast that grain refinement and modification are also not required. However, it is found that treatment of the liquid alloy by a Rotary Degassing Unit is effective in removing non-metallic inclusions, reducing sedimentation of hard iron–manganese inclusions as well as lowering gas content. Increasingly diecasters are using Rotary Degassing in the transfer ladle for this reason. Some diecasters find that modification with strontium is also valuable.

Metal treatment

It is important to keep melting and holding furnaces clean to prevent hard inclusions of corundum from entering the castings. Regular use of a cleaning flux, COVERAL GR2220 or COVERAL 88, is recommended. Bulk melting is carried out under a covering/drossing flux to ensure minimum metal loss. COVERAL GR2220 granular flux (0.5 kg/square metre) or COVERAL 5F powder flux (1 kg/square metre) is used forming a complete cover, adding half early and the rest at final melt down.

The metal is poured into the transfer ladle and degassed using the Mobile Degassing Unit or the fixed Foundry Degassing Unit (FDU), modification using SrAl piglets may also be done at the same time. Note that sodium modification is not suitable since the metal may be held for a long time in the holding furnace before use.

In the holding furnace, COVERAL 72 low temperature flux can be used as a cover to reduce metal loss. Scatter 0.25–0.5% onto the metal surface and rabble gently until the exotherm develops. Push aside or remove before taking ladles.

Die coating

Refractory die coatings are not used for pressure diecasting, since high heat transfer is needed to cool the casting quickly and achieve fast casting cycle times. However, the die must be sprayed between each shot with a lubricant.

The conditions under which a die lubricant must work are particularly severe. The die cycles at a high temperature (250–300°C) and at regular intervals molten metal is injected at great pressure. All the alloys cast commercially will attack and weld to steel dies. The lubricant must protect the expensive die from direct metallurgical attack and erosion as well as lubricate slides, cores, ejector pins etc. to prevent them seizing at the high operating temperature. The coating also has a cooling function. Only occasionally can a single lubricant be relied upon to perform all functions satisfactorily, so pressure diecasting lubricants are usually divided into two types:

> Mechanism lubricants
> Die face lubricants

Specialist lubricant suppliers such as Chem-Trend supply a wide range of products for all types of pressure diecasting.

Plunger and mechanism lubricants

Chem-Trend has an extensive range of liquid plunger lubricants, both water and oil based in graphited and non-graphited versions with viscosity designed to meet particular applications. Chem-Trend also produce a full range of solid plunger lubricants from beads through to powders. These materials are manufactured both graphited and non-graphited. Solid lubricants work as well as oil lubricants and are cleaner environmentally. Solid lubricants are injected into the cold chamber with the Power Lube PL300 Applicator in controlled amounts every shot at the rate chosen by the operator, usually 0.5–2.5 grams depending on the size of the plunger tip and the casting weight.

Die face lubricants

There is a wide range of Chem-Trend SAFETY-LUBE products which can be tailored to particular applications, for example:

> Aluminium   SL–7562 for medium to heavy duty aluminium castings
>   SL–7577 for heavy duty aluminium castings

These are concentrated emulsions of synthetic oils and lubricity additives in water designed to enhance lubrication of ejector pins and sliding cores together with anti-soldering and parting additives. The concentrates are diluted with water at a starting dilution of 1:80 or 1:100 depending on the complexity of the parts. Dilution can be extended further after the natural burnished coating provided by the lubricant has been established. This normally takes place after about 2 hours. The lubricant is applied by mechanised spray equipment specifically designed for each die set or by a hand spray gun.

Zinc	SL–7552 water based for all zinc castings. Use a starting dilution of 1:30
	SL–7578 solvent based for the miniature zinc diecasting process. Ready for use
Magnesium	RDL–2649 water based

Chapter 10

Low pressure and gravity diecasting

Low pressure diecasting

The principle of the process is shown in Fig. 10.1. A metal die is mounted above a sealed furnace containing molten metal. A refractory-lined tube, called a riser tube or stalk, extends from the bottom of the die into the molten metal. When air is introduced into the furnace under low pressure (15–100 kPa, 2–15 psi), the molten metal rises up the tube to enter the die cavity with low turbulence, the air in the die escaping through vents and the parting lines of the die. When the metal has solidified, the air pressure is released allowing the still-molten metal in the riser tube to fall back into the furnace. After a further cooling time the die is opened and the casting extracted.

The process is capable of making high quality castings. With correct die design, directional freezing of the casting is achieved so eliminating the need for risers, the casting being filled and fed from the bottom. Because there is usually only one ingate and no feeders, casting yield is exceptionally high, generally over 90%. Good dimensional accuracy and surface finish are possible and complex castings can be made using sand cores.

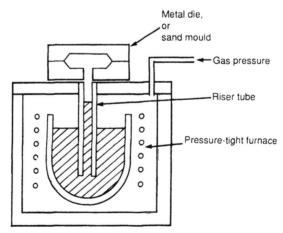

Figure 10.1 *The prinoiple of a low pressure diecasting machine. (From Campbell, J. (1991)* Castings. *Butterworth-Heinemann, reproduced by permission of the publishers.)*

Typical applications

Aluminium automotive parts: wheels, cylinder heads, blocks, manifolds and housings
Critical aerospace castings
Electric motor housings
Domestic kitchenware such as pressure cookers

Large castings up to 150 kg (Al) can be made but can only be justified in special cases because of high die costs.

The design of a typical low pressure diecasting machine is shown in Fig. 10.2. The molten metal is contained in a crucible heated by electrical resistance windings. The capacity of the furnace is usually sufficient to make around 10 castings before refilling is necessary. The crucible can be topped up with molten metal as necessary via a filler port. The whole furnace is contained in a pressure vessel sealed with a gasket by a top plate. The riser tube or stalk is suspended from the top plate by a riser cap or nozzle. The upper part of the riser tube is heated by a gas burner to prevent the liquid metal being cooled by the water-cooled lower die-half. The riser tube is immersed in the molten metal nearly to the bottom of the crucible. The riser

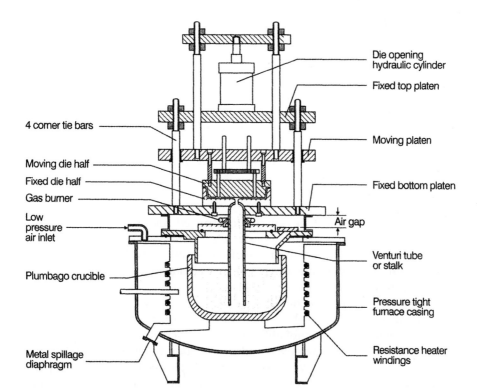

Figure 10.2 *Low pressure diecasting machine construction. (P.A. Bryan,* The British Foundryman, **65**, *Nov. 1972, p. 401.*

tube may be made of ceramic but is more commonly made of steel or cast iron coated with a refractory wash to prevent attack by the molten aluminium. A steel or cast iron stalk, properly coated with the correct refractory dressing at regular and frequent intervals (usually once per shift), has a life of around 6 months. An INSURAL tube may also be used to line the stalk, see below.

The lower fixed die-half, mounted on a base plate, is fitted to the furnace top plate. The base plate carries 4 corner tie bars on which slides the moving platen carrying the upper die-half. A hydraulic cylinder mounted on a fixed top platen opens and closes the die. The die itself may carry hydraulically operated core pulls. The whole unit is designed in such a way that the top plate carrying the die and its opening mechanism can be swung away to permit easy access to the riser tube and furnace.

When the holding furnace is at temperature, a little above the melting point of the alloy being used, it is filled by way of the filler port which is then sealed. When the metal temperature has stabilised to the required value and the die has been preheated to its operating temperature (250–400°C) and closed, the inlet valve is opened and dry compressed air is allowed to fill the sealed furnace to a controlled pressure causing the aluminium to rise in the transfer tube and fill the die. With the furnace remaining under pressure, the casting solidifies quickly, the direction of freezing following a downward path with the sprue section being the last to solidify. When the metal in the nozzle has frozen, the pressure is released allowing the still-molten metal in the riser tube to fall back into the furnace. A further short time is allowed to ensure complete solidification of the casting, the die is then opened and the casting released into the upper die-half from which it is retrieved, usually mechanically. Once the sequence has been established, it can be controlled automatically using temperature and pressure controllers and timers enabling one operator to supervise more than one diecasting machine.

Die design

Low pressure dies must be solidly constructed and are normally manufactured from cast iron. Wearing parts, retractable cores and areas of the die where heat is concentrated are made of H13 or an equivalent tool steel. The dies are built with the same degree of precision as those used for pressure diecasting to give positive and reliable operation.

The thermal balance of the die has to be carefully evaluated since the casting must freeze from the extremities back to the riser tube to avoid shrinkage defects. This is probably the most difficult aspect of low pressure diecasting and much experience and cooperation between die maker and foundry is needed to achieve success. It may be necessary to force-cool parts of the die, by air or water. A new die rarely operates correctly first time and a long period of sampling and modification may be needed before the die can be operated at maximum efficiency.

Dies must be coated with a suitable dressing to avoid welding of the molten aluminium to the die and to control the rate of heat extraction. The DYCOTE dressings are the same as those used for gravity diecasting, see below. Application is graduated from very thin, where maximum chilling is required, e.g. isolated bosses, to quite thick on thin metal sections. The coatings are applied by spray with the die at a temperature of 120–200°C so that the coating dries quickly without running. The die must then be reheated above working temperature before casting metal. The frequency with which the coating must be touched up or reapplied must be found by experience, the aim must be to recoat only once per shift. It is very important that particular care is taken in cleaning dies regularly, say once per week, by "shot"-blasting using glass beads, otherwise tolerance and surface finish of the castings will soon deteriorate.

Die life is normally around 30 000–50 000 "shots".

Cores

Sand cores made by any of the usual processes – shell, cold box, hot box etc. – can be used. As with any casting process, cores must be permeable and provision made for venting core gases to avoid gas defects in the castings. Where complex coring is needed, such as for cylinder heads, it is usual to preassemble a core package which can quickly be inserted into the die to avoid slowing down the casting cycle.

Cycle time

The process is rather slow since the die must be filled slowly to avoid turbulence and air entrapment. The casting must solidify progressively from the extremities back to the gate and the gate must solidify before the pressure can be released. Cycle times are similar to gravity diecasting, about 9 shots per hour are typical for a component such as a cylinder head.

Alloys cast

The commonly used aluminium alloys include:

Al–Si5Cu3 (LM4, US 319)
Al–Si12   (LM6, US 413)
Al–Si7Mg   (LM25, US A356)

It is usual to fully treat the metal by degassing, grain refining and modifying (see Chapter 6).

Casting quality

The process has, in principle, all the features necessary to produce castings of high quality, both metallurgically and dimensionally:

> The metal is drawn from the bottom of a bath of molten alloy, avoiding the contaminated surface layer,
> The mould is filled gently without turbulence so avoiding oxide entrapment,
> Solidification is directional, enabling constant feeding of the casting by maintenance of pressure from below,
> Mechanical casting extraction avoids damage to the die so that dimensional accuracy should be maintained throughout the life of the die.

In practice, there are still many problems which may affect quality. Although metal is drawn from below the surface of the melt, oxide inclusions may still be present due to the turbulence arising when the furnace is topped up from the transfer ladle. The density of aluminium oxide is close to that of the metal itself so oxide inclusions may not all float to the surface. Further turbulence is introduced in the furnace by the fall of the metal in the riser tube each time a casting is made and the pressure released. Casting quality can be improved by fitting a SIVEX FC ceramic foam filter in the sprue to prevent inclusions from entering the casting.

Casting tolerances in low pressure diecasting are similar to those of gravity diecasting. Variation in the amount of die coating applied to the die is one of the main reasons for dimensional inaccuracies. Die distortion over time will also affect dimensions of the castings.

Although low pressure diecasting is rather slow and expensive, its ability to produce high quality castings reliably has made it a preferred method for castings such as high quality car wheels, and many cylinder heads and manifolds are cast using the process.

Use of INSURAL refractory

The durability, non-wetting and insulating properties of INSURAL (Chapter 5) make it valuable for use in low pressure diecasting applications.

The stalk or riser tube requires special protection or the molten aluminium will attack and dissolve the steel or cast iron, contaminating the casting alloy. Continuous metal movement inside the tube places severe strains on any protective coating. Both the inside and outside of the stalk should be given a carefully applied coating of DYCOTE 976 dressing. This coating has been formulated to have extremely good resistance to erosion and a tenacious bond.

Better still is to line the inside of the stalk with an INSURAL tube, Fig. 10.3. Not only does this protect the steel tube, but by keeping the liquid

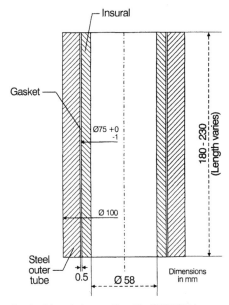

Figure 10.3 *Lining the inside of the stalk with INSURAL.*

metal hot, it eliminates the need to heat the top of the stalk with external gas burners. The riser cap or nozzle may also be lined with INSURAL, Fig. 10.4. This allows the die to be water cooled effectively while still maintaining metal temperature in the nozzle to permit good filling and feeding of the casting.

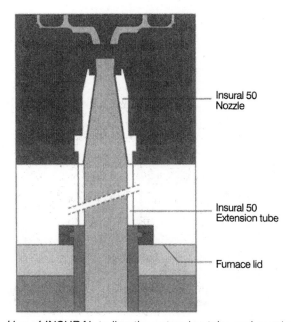

Figure 10.4 *Use of INSURAL to line the extension tube and nozzle.*

Gravity diecasting

The molten metal is poured under gravity into a refractory-coated permanent mould or die. The technique is sometimes known as "permanent mould" or "chill" casting. The dies are made of a fine-grained, pearlitic cast iron or a low alloy steel. Simple retractable cores may be made of high grade alloy steel, but resin bonded sand cores are used to produce complex internal shapes.

With the exception of pressure diecasting, the process is the most widely used of the aluminium casting methods due to its inherent simplicity and the metallurgical quality and complexity of castings that can be made. The process is used for castings made in numbers from 1000 to more than 100 000 per year, for example manifolds, cylinder heads, water pump housings etc. Casting size ranges from less than 1 kg to over 50 kg.

Figure 10.5 *A manual gravity die set.*

Diecasting machines range from simple, hand-operated rack and pinion die sets, Fig. 10.5, manually poured, which can profitably make runs of as few as 1000 parts, to carousel machines having several dies automatically operated with cores placed by robot and poured automatically with mould and metal temperatures, casting speed etc. being closely controlled. Unlike pressure diecasting, there is no necessity for powerful locking forces to hold the die halves together while metal is cast, since the pressures involved are relatively small. Complex dies are usually opened and closed hydraulically.

The dies are coated with a refractory-based coating (see below). The type of die coating used and its application are critical. The coating reduces the heat transfer to the die so that cooling rates are faster than in sand moulds but slow enough for complex castings to be filled satisfactorily, Table 10.1. Different coatings are available which allow faster or slower cooling as the shape and section of the casting requires. The time before the casting can be extracted from the die may vary from 4–10 minutes depending on the type of casting. The process is therefore relatively slow. To achieve reasonable output rates, a manual operator will usually operate 2–4 die sets in sequence, having a total output of 30–60 castings per hour.

Table 10.1   Solidification times of Al castings made by different processes

Casting process	Mould material	Solidification time (seconds)
Permanent mould	Steel	47
Core	Silica sand	175
	Zircon sand	80
Disamatic	Silica/clay	85

(From Hansen P.N., Rasmussen N.W., Andersen U. & M. *AFS Trans.*, **104**, 1996, p. 873)

Automatic carousel machines may have from 4 to 6 stations with multiple die sets allowing production rates of around 1 casting per minute (200 000 to 300 000 parts per year).

Dies usually run at a temperature of 300–350°C with the die face temperature higher than the back of the die block. This can lead to distortion and bowing of the die which may allow flash and even run-outs to occur. The core packages must fit the dies with sufficient clearance to avoid damage to the cores and allowance must be made for some variation in the thickness of the die coating. Differential thermal expansion of the die, together with the variable thickness of the coating, place limits on the dimensional accuracy of castings made by the process and metal sections are limited to about 4 mm because of dimensional factors and because the chilling effect of the die limits the ability of the metal to run in thin sections.

Some control of die temperature can be achieved by using dedicated gas burners controlled by thermocouples inserted into key areas of the die. The dies can then be externally heated in preparation for application of the coating and to hold temperature during breaks or downtime.

Dies must be vented to allow air to escape as the casting is poured. Core prints always have some clearance which acts as a vent, as does the mould parting which, due to thermal expansion of the die, is always imperfect. Risers, needed to feed metal shrinkage, also provide venting for the mould cavity. If additional venting is needed, slots up to 0.3 mm deep may be cut on the parting face of the die.

Dies must be designed to allow easy removal of the finished casting. Manually operated dies are often designed to allow a hand-held lever to be used to free the casting from the cavity without damage to a critical area of the casting. Automatic diecasting machines usually have a dedicated unloader to remove the casting. The die may be designed with several moving parts to allow potentially difficult areas of the die to be stripped from the casting while the casting it is still being firmly held by the remainder of the die.

Die lives are typically around 30 000–50 000 shots. Dies are increasingly being designed with the aid of computer programs which simulate the solidification process. Their use can save weeks of trial and error, significantly reducing the time and cost of bringing a new casting into production.

Running and feeding

The gravity diecasting process is capable of making castings of high metallurgical quality. Metallurgical structures and properties benefit from the chilling achieved by gravity diecasting, Table 10.1. The design of the running and feeding systems has a major effect on both quality and yield of castings. The objective of running the metal into the cavity is to make entry as smooth as possible, avoiding turbulence which would introduce oxide film defects. The speed of filling is a compromise between getting the metal to the furthest point of the die quickly enough to avoid misruns, and pouring slowly enough to avoid turbulence. Bottom gating is the traditional way of achieving turbulence-free filling but bottom gated castings have poor yields and incorrect thermal gradients.

Feeders need to be at the top of the casting and must be the last masses to freeze, but optimum directional solidification is not easy to achieve when the metal is bottom gated and has to traverse the die cavity and any cored passages before reaching the feeders. One way of improving feeding is to tilt the die when pouring in such a way that the last metal is poured into the feeder. Special tilting machines are available to allow this to be done.

The widespread use of ceramic foam filters and insulating feeders has greatly improved the quality of gravity castings in recent years (see Chapter 8). A recent development has been the use of direct pouring methods, either through a filter fitted directly in the die or through a combined filter/feeder. This greatly improves yield, achieves turbulence-free filling and also feeds the casting with the hottest metal (see page 107).

Figure 8.8 shows a 3.4 kg thin-walled casting made in LM6 (Al–Si12) top poured through a ceramic foam filter in a single down-sprue. This method enabled the casting to be made successfully with high yield and without the need for vacuum impregnation.

Melting and metal treatment

The usual alloys cast by gravity diecasting are:

LM4   (Al–Si5Cu3, US 319)
LM6   (Al–Si12, US 413)
LM25 (Al–Si7Mg, US A356)

LM6, being the eutectic alloy, has the best fluidity and is good for thin section castings but its machining characteristics are poor. LM25 has good fluidity and good machining properties, it can be heat treated and is corrosion resistant. LM4 has somewhat lower fluidity due to its low Si content and has good machinability.

It is advantageous for a foundry to use the minimum number of different alloys, allowing bulk melting to be used for the majority of the castings. Since gravity diecasting is capable of making high quality castings and is used for critical castings such as cylinder heads, hydraulic castings etc., it is usual to fully treat the metal by degassing, grain refining and modifying (see Chapter 6).

Cores

Sand cores made by any of the usual processes – shell, cold box, hot box etc. – can be used (Chapter 13). As with any casting process, cores must be permeable and provision made for venting core gases to avoid gas defects in the castings. Where complex coring is needed, such as for cylinder heads, it is usual to preassemble a core package which can quickly be inserted into the die to avoid slowing down the casting cycle.

Die coatings for gravity and low pressure diecasting

A gravity or low pressure diecasting must be metallurgically sound, have good surface finish and be easily and rapidly produced. To achieve this, the die must be coated. The die coating serves several functions, it should:

Protect the accurately machined die face
Ease the release of the casting from the die
Control heat flow from the metal to the die

Provide good surface finish to the casting
Lubricate all moving parts of the die

The coating should also prevent build-up of residues on die faces and be free from excessive fumes.

Control of heat flow

The control of heat flow from the metal to the die is the most important property of the coating, allowing control over filling the thin sections and the solidification of the casting. The thermal insulation properties of the applied coating layer are determined by three key factors, namely:

Coating composition
Layer thickness
Coating layer porosity

Coating composition

Permanent mould coatings are typically formulated using water as carrier, a high temperature binder (normally sodium silicate), and a refractory filler or blend of fillers. Table 10.2 lists the range of DYCOTE dressings available from Foseco. There are two categories of coatings:

1 Insulating   Contain blends of insulating minerals such as talc, mica, kieselguhr, titanium dioxide, alumina etc. (DYCOTE 34, 140)
2 Lubricating Based on colloidal graphite or boron nitride to aid release of the casting (DYCOTE 36, GL1200)

At the normal casting temperatures of aluminium alloys the refractoriness of the fillers is sufficient to ensure that chemical change does not occur and their function is essentially physical. Thermal conductivity and granulometry are key properties and the diecaster needs to consider the most important aspect of the particular casting to be produced when selecting the die coating or coating combination, e.g. high insulation to avoid misruns, lubrication, smooth surface finish etc.

It is generally accepted that a rough coating surface provides maximum insulation through the formation of an air gap at the metal/coating interface during pouring. The air gap arises from the high surface tension of the liquid alloy. The low metal/refractory contact area of a rough coating surface significantly reduces heat transfer and consequently aids metal flow. In addition, coarseness and angularity of the refractory particles enhances metal flow by continuously rupturing the oxide skin on the molten metal surface as it flows over the mould. Conversely a smooth coating based on

Table 10.2 DYCOTE insulating dressing for gravity and low pressure diecasting

	Description	Dilution	Characteristics	Typical Components
DYCOTE 6	Light grey paste	3–4 vol. soft water to 1 vol. paste	A general purpose dressing gives a medium finish.	Mica, talc, sodium silicate
DYCOTE 32 ESS	Pastel pink slurry	3–5 vol. soft water to 1 vol. slurry	A medium grade, fairly soft dressing. Contains iron oxide. Used mainly on wheels made by low pressure diecasting.	Calcium carbonate, iron oxide, sodium silicate
DYCOTE 34 and DYCOTE 34 ESS[1]	Grey paste	3–5 vol. soft water to 1 vol. paste	Provides a medium to rough dressing having high insulation values. Very suitable therefore for thin casting sections and large flat surfaces.	Mica, talc, carborundum grit, sodium silicate
DYCOTE 39 and DYCOTE 39 ESS[1]	White paste	3–5 vol. soft water to 1 vol. paste	A superfine dressing of very high refractoriness. Suitable for medium to thick casting sections. Also useful for slush casting.	Titanium dioxide, sodium silicate
DYCOTE 140 and DYCOTE 140 ESS[1]	Off white paste	3–5 vol. soft water to 1 vol. paste	The standard, general purpose dressing used extensively in all sectors. It gives a medium to fine finish.	Mica, talc, sodium silicate
DYCOTE 36	Black paste	3–5 vol. of soft water to 1 vol. paste.	Has some lubricating properties due to its graphite content. Hard wearing and most suitable for cast iron and copper base alloys.	Mica, talc, graphite, sodium silicate
DYCOTE 976	Dark grey slurry	Ready for use	A specially formulated protective coating for the stalks in low pressure diecasting.	Graphite, alumina, boric acid, PVA Binder
DYCOTE GR 8700	Red brown slurry	3–5 vol. of soft water to 1 vol. paste	A primer coating for use on all the die face before insulating coating applied. Improves life of coating on die. Touch up more successful.	Iron, oxide, zirconium silicate, sodium silicate
DYCOTE GL 1200	White paste	10–20 vol. of soft water to 1 vol. paste	A lubricating coating used with a base coat of DYCOTE GR 8700 to ensure clean strip and good surface finish	Boron nitride

Note: The suffix ESS indicates an extra strongly bonded version of the dressing.

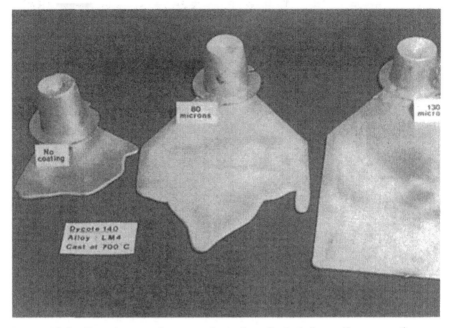

Figure 10.6   *Test plate casting to evaluate the effect of die coating properties on metal flowability.*

fine fillers provides a larger contact surface area and higher heat transfer rate.

Figure 10.6 shows a simple test plate casting used to illustrate the effect of coating insulation and granulometry on metal flowability, in this case showing the effect of varying DYCOTE 140 thickness.

Coating layer porosity

Insulation of the coating layer is dependent not only on thickness but also on inherent porosity. This property in turn is influenced greatly by application method and conditions – the degree of pore formation essentially being determined by the rate of evaporation of the water carrier on contact with the die. Although die dressings are occasionally applied by brushing, especially in running, gating and riser areas, the most common method is to apply by spraying. Application of the dressing at a mould temperature of around 170–200°C usually gives optimum results, although coating consistency and spray equipment also influence the quality of the deposited coating layer and slight deviations from this temperature may be necessary.

Higher mould temperatures usually result in greater layer porosity due to a more vigorous reaction between the water carrier and the hot die. Although insulation is accordingly higher, coating adherence and durability

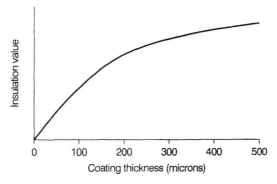

Figure 10.7 *Relationship between coating thickness and insulation value.*

are adversely affected. Excessive die temperatures can lead to violent kick-back, preventing the coating from adhering to the surface at all. Conversely, lower mould temperatures result in a denser, less insulating layer, albeit of greater durability. Excessively low temperatures can lead to "puddling" of the coating resulting in cracking and blistering defects as the die is heated to operating temperature and trapped moisture ruptures the binder film as it escapes. Porosity in the coating layer actually decreases the greater the distance from the mould surface due to progressively less vigorous water evaporation during coating application.

Figure 10.7 shows a typical graph of insulation as a function of thickness indicating that a 300 micron coating layer would not be significantly more insulating than one of 200 microns. As heavy layers over 300 microns also tend to be more susceptible to flaking from the die, the optimum target operating range for thickness is 150–250 microns.

Coating the die

Selection of the correct coating for a particular casting only results in high quality castings if the die coating operation itself is controlled and consistent. Often, however, this is one of the most abused and variable aspects of the entire casting process. Careful attention to die preparation, coating preparation and application, and the type of coating equipment utilised, can yield significant quality and productivity benefits.

As with all coating applications, careful preparation of the substrate is critical to the ultimate performance of the die dressing. New dies need to be thoroughly cleaned to dispel any lubricants or rust preventive treatments, while dies already in service must have the old coating layer completely removed. Traditional die cleaning methods include wire-brushing and blasting with either sand, metal shot or grit. Excessive wear and tear of the die surface has caused diecasters and suppliers alike to look at non-destructive cleaning methods including wet chemical methods and resin media as used in the aircraft industry to remove old paint. Neither of these options has yet been developed commercially due to environmental and cost concerns respectively. Die cleaning through blasting with carbon dioxide pellets is now becoming popular, particularly in North America.

After cleaning, the die should be heated uniformly to around 300–350°C. This is usually achieved through the application of gas burners appropriately positioned inside and around the die although temperature variations over the surface of the mould are inevitable with this method. Many foundries are now preheating dies in dedicated ovens under tightly controlled conditions in order to achieve an extremely even, uniform temperature profile across the die. Dies are normally cooled to around 180°C for coating application. A primer, either a water/surfactant solution or very dilute version of the regular coating, is often used to promote enhanced adhesion of the main coat.

Coating preparation and application

Proper control of coating properties at application is critical. Traditionally operators have used approximate dilution ratios when preparing the coating for use but control methods such as baumé and density are now being more widely adopted. Many foundries now monitor applied coating thickness during spraying of the die by using compact portable gauges operating on the magnetic and eddy current principle. Mould temperature during spraying is critical to the behaviour of the coating in service and consequently an increasing number of foundries now measure die temperature using either optical pyrometers, computer-linked thermocouples, or simple contact thermometers.

Application equipment

The most popular application technique still used today involves a simple venturi system comprising a small container, copper piping and compressed air line. While this system is capable of giving excellent results, it does lack control and can result in a great deal of spattering, giving rise to an uneven and irregular coating layer. More foundries are now adopting airless spray systems and purpose-designed spray guns which deliver a fine, uniform coating "mist" via an adjustable fan width. Equipment has also been developed to provide not only agitation in the feed container but also recirculation of material in the feed hoses themselves whilst the gun is not in use in order to help prevent separation of liquid/solids at all times (Figure 10.8).

Computer simulation

Computer simulation of die filling and solidification profiles is becoming an extremely powerful tool for foundries in the design and optimisation of process parameters for new or prototype castings. Today a number of

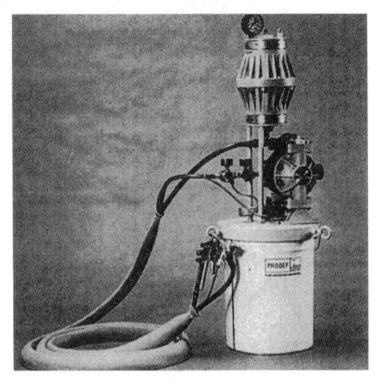

Figure 10.8 *Spray gun for coating dies.*

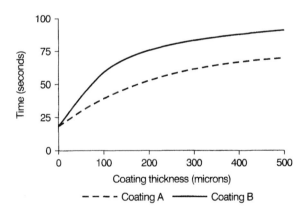

Figure 10.9 *Effect of coating thickness and type on casting solidification time.*

programs are commercially available to assist diecasters to predict how dies
will behave in service before they are built. Various design and process
parameters can be tested and validated before tooling is built, thereby
improving the diecasters' "right first time" capability and avoiding time-
consuming and costly start-up trials. The impact of coating type and

application on die filling, solidification and casting and mould temperature profiles is just one process parameter which can be assessed.

Figure 10.9 shows a computer simulation comparison of casting solidification times for two coatings. A and B; A having greater insulation properties than B. The comparison shows the effect of coating thickness on solidification time. Coating B would have to be applied at more than 0.3 mm to have the same insulation effect as coating A at 0.1 mm. Data of this kind can greatly assist the diecaster in understanding the likely effect of different coating practices and help in process optimisation.

Chapter 11

Sand casting processes

Introduction

Gravity and low pressure diecasting techniques are inherently slow because the die is effectively out of use during the time that the casting is solidifying. When the casting is removed, the die must be reloaded with cores and closed before the next casting is poured. This may take 10 minutes or more for a large, complex casting such as a cylinder block. In order to achieve the high production rates required by automotive manufacturers, expensive multiple die sets are required. With sand casting, by contrast, productivity is not affected by the solidification time but only by the rate at which the sand mould/core package can be produced. With automatic green sand moulding and cold-box coremaking, mould/core packages can be made and assembled in times which enable high production rates to be achieved without the need for the costly multiple tooling sets needed by diecasting. Sand casting, which includes green sand, core assembly processes and the Lost Foam process, accounts for around 12–15% of all aluminium castings, Table 11.1.

During the last few years, the use of Lost Foam casting has increased significantly, but is still probably less than 10% of the total.

Table 11.1 Share of Al castings market held by various processes (Rasmussen N.W., *The Foundryman*, **87**, 1994, p. 63)

Casting method	*% by weight (1991)*	
	USA	*Germany*
High pressure diecasting	60–65	55
Low pressure and gravity diecasting	20	32
Sand	10	12
Lost Foam	2–5	–
Others	<5	<1

Green sand

With the exception of pressure diecasting, which is limited to uncored castings, green sand moulding is the most productive casting process. Moulding rates up to 6 or 7 moulds per minute from one set of tooling are possible with automatic machines such as the DISA flaskless moulding machine. Clay binders have low cost and reuse of the sand is possible. It is surprising then that green sand casting of aluminium alloys is not used more widely. Probably the high capital cost of green sand plant compared with the simplicity of gravity diecasting accounts for its relative unpopularity. The growth in the use of aluminium for automotive applications over the past few years has stimulated a renewed interest in the automatic green sand method.

The slower rate of cooling of sand castings compared with diecastings (Table 10.1) leads to smaller temperature gradients and a wider solidification time range, making feeding more difficult. The grain structure of slower-cooled sand cast aluminium is usually regarded as inferior to that of chill cast alloys. To counter this, moulding sand based on magnetite ore has been proposed to increase the chilling effect.

There has also been some concern that aluminium alloys may pick up hydrogen from moisture in the green sand, but it is not a problem if the sand system is correctly controlled to avoid excessive water. Unlike green sand used for iron casting, it is not necessary to have coal dust in the sand to improve casting surface finish. There are some foundries which use the same sand system (with coal dust) for both iron and aluminium casting without any harmful effect. See Chapter 12 for details of sands and sand systems.

To make high integrity castings using green sand, the liquid alloy must be carefully treated before casting to remove hydrogen and oxides from the melt and the alloy must be grain refined and modified as described in Chapter 6.

As with other casting processes, there is a danger of introducing oxide films into the metal during molten metal transfer, pouring into the mould and within the running system if turbulence occurs. The requirement for fast casting rates to justify the high capital cost of green sand moulding machines and the associated sand plant conflicts with the need for low turbulence during mould filling. One foundry uses a computer-controlled electromagnetic pump to fill vertically parted DISA moulds without turbulence. A special shut-off device is needed to allow the mould to be moved immediately after filling so that high production rates are possible. Other foundries use carefully designed down-sprues to reduce splashing in the empty gating channels during filling of the gating system. Gating designs with low Reynolds number are claimed to ensure calm filling of the mould. The Reynolds number of a hydraulic channel determines the onset of turbulence. Low velocity filling with flat runner designs (which ensure low Reynolds number) are considered to be best.

The simplest method of filling moulds rapidly and without introducing oxide defects into the casting is by using ceramic foam filters in the running system (see Chapter 8). These have the twin effects of removing already entrained oxide and of reducing turbulence downstream of the filter. Top pouring through a filter is an efficient way of filling sand moulds, giving high yield and excellent mechanical properties, Fig. 2.1.

To produce castings free from shrinkage porosity, directional solidification must be encouraged, with the use of feeders to supply liquid metal to the last sections of the casting to solidify (see Chapter 7). Simulation techniques are increasingly being used to design running and feeding systems.

Moulding machines

The moulding machine must compact the green sand evenly around the pattern to give the mould sufficient strength to resist erosion while liquid metal is poured and to withstand the metallostatic pressure exerted on the mould when full. Green sand does not move easily under compression forces alone so that achieving uniform mould strength over a pattern of complex shape by simply squeezing is not possible. Various combinations of jolting and vibrating with squeeze applied simultaneously or sequentially have been used to produce uniform strength moulds at up to 400 moulds per hour. Green sand moulds have traditionally been made in steel flasks, but flaskless moulding is widely used for smaller castings.

Moulding in flasks

The basic principles used are:

Jolt squeeze

A pattern plate carrying the pattern surrounded by a moulding flask is fitted onto a jolt piston. Green sand from a hopper above the machine fills the moulding box loosely. The assembly is raised by the pneumatic jolt cylinder and allowed to fall against a stop. The jolt action is repeated a preset number of times causing the sand to be compacted to some extent, the density being highest nearest the pattern plate. Additional compaction is provided by a pneumatic or hydraulic squeeze plate, Fig 11.1a. More uniform squeeze can be achieved by using compensating squeeze heads, Fig. 11.1b, which compact the sand more uniformly even though the depth of sand over the pattern varies.

Vibration squeeze

Instead of a jolt table, the pattern and flask are vibrated before and during squeezing by a multi-ram head.

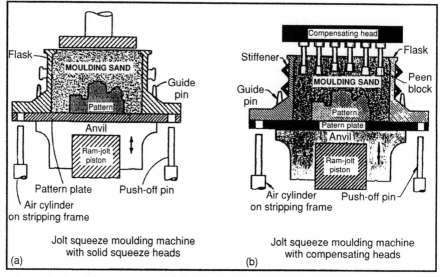

Figure 11.1 *Jolt squeeze moulding machine. (a) with solid squeeze head, (b) with compensating heads. (Sixth Report of Institute Working Group T30, Mould and Core Production,* Foundryman, *Feb. 1996.*)

Shoot squeeze

A shoot head above the flask contains loose green sand. Opening a shoot valve allows compressed air stored in a reservoir to travel through the shoot head causing a column of sand to be shot into the flask where the kinetic energy of the rapidly moving sand compacts it against the pattern. A hydraulic cylinder then squeezes the mould to complete the compaction.

Impulse compaction

Sand is compacted into the flask by means of a pressure wave generated by explosion or by compressed air. Compaction is greatest near the mould surface. Squeeze pressure may then be applied to compress the back of the mould. Deep pattern draws are possible with this method.

Vacuum squeeze

A partial vacuum is created in the flask surrounding the pattern. A metered amount of sand is released into the chamber where the vacuum accelerates the sand which impacts onto the pattern causing compaction. A multi-ram head provides high pressure squeeze to complete the compaction of the mould. The system is suitable for large moulds.

Flaskless moulding

Horizontally parted (matchplate moulding)

A matchplate is a pattern plate with patterns for both cope and drag mounted on opposite faces of the plate. Both cope and drag halves of the mould are alternately filled with prepared sand before being brought together on either side of the matchplate pattern for the mould halves to be formed by high pressure squeeze with simultaneous vibration to compact the sand. The mould halves are separated to allow the matchplate to be withdrawn and cope and drag to be closed. The completed mould is finally pushed out of the machine onto a shuttle conveyor with the bottom of the mould resting on a bottom board to facilitate progress along the mould conveyor. Hunter matchplate moulding machines are available with mould sizes from 355 × 483 × 40/114 mm up to 762 × 813 × 305/279 mm. Moulds can be made at up to 200/hr on the smaller machines and 120/hr on the largest.

Vertically parted moulding

The Disamatic flaskless moulding machine introduced in the late 1960s (now supplied by Georg Fischer Disa) revolutionised green sand moulding, allowing high precision moulds to be made at up to 350 moulds/hr. The method of operation is shown in Fig. 11.2. One pattern half is fitted onto the end of a hydraulically operated squeeze piston with the other pattern half

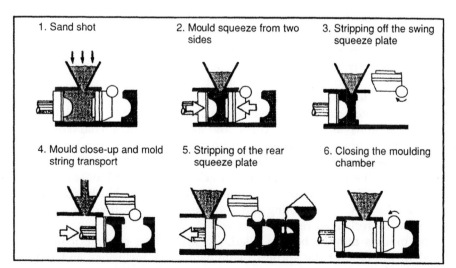

Figure 11.2 *Vertically parted flaskless moulding, the Disamatic machine. (Sixth Report of Institute Working Group T30, Mould and Core Production,* Foundryman, *Feb. 1996.)*

fitted to a swing plate, so called because of its ability to move and swing away from the completed mould. Sand from a supply hopper above the machine is blown into the moulding chamber by means of a variable pressure, compressed air supply stored in a nearby air receiver. Vacuum can be applied to the moulding chamber to vent air and assist in drawing sand into deep pattern recesses.

Both halves of the pattern are hydraulically squeezed together to compress the sand block. As the swing plate moves away, the piston pushes the new mould to join ones previously made, to form a continuous mould string. Mould sizes available are from 500 mm × 400 mm × 315 mm on the smallest 2110 model, up to 950 mm × 800 mm × 635 mm on the largest model manufactured, the 2070. Flexibility is available through variable mould output, variable mould thickness, fast pattern change and core placing options. Varying degrees of control sophistication are provided dependent on the model. Cores can be placed in the mould using a mechanised core placer.

There are many variations on the moulding principles described above – see Sixth Report of Institute of British Foundrymen Working Group T30 (*Foundryman*, Supplement, Feb. 1996, pp. 3–30) from which some of the above information has been taken.

Core assembly sand processes

Automatic green sand lines require expensive moulding machines and sand treatment plant. They are somewhat inflexible since the size of the moulding plate determines the type of casting that can be economically made. Clearly it is not possible to make a casting larger than the plate size. Smaller castings may also be uneconomic since the true capacity of the expensive machine may not be fully utilised. One solution, particularly for complex, highly cored castings, is to use a core assembly technique.

Modern cold-box coremaking methods are now so fast that high production rates can be achieved by using coremaking techniques for forming moulds as well as cores. By hardening in a cold core box, dimensionally precise cores and mould sections can be produced. The parts can be subsequently assembled accurately, manually or by robot, to make a package which is placed on a conveyor track to be cast. Components of great complexity such as manifolds, cylinder heads, cylinder blocks etc. can be made in this way. After cooling and knocking out the casting, the sand can be thermally reclaimed ready for reuse.

The moulds may be bottom filled, usually with a ceramic foam filter fitted in the base of the sprue to eliminate oxide inclusions and reduce turbulence in the running system. Alternatively, direct top pouring through a KALPUR unit can be used as has been described in Chapter 8, Figs 8.6 and 8.7.

The Cosworth process

The Cosworth process uses a liquid metal pump to fill the mould from below with minimum turbulence, in a similar manner to low pressure diecasting, Fig. 11.3. Cores are made from zircon sand, which has a more uniform and lower thermal expansion rate than silica sand, and it is claimed that dimensional errors due to expansion are reduced. The chilling effect of zircon also helps to give fine-grained casting structures (see Table 10.1). Molten aluminium alloy is held in a large (10 tonne) electrically heated radiant roof holding furnace, topped up as required from a melting furnace. The bath is maintained under an inert atmosphere to limit the amount of oxidation and continuously degassed with argon at the charging point. The large capacity of the holding furnace ensures that oxide inclusions formed during the melting and topping-up operations have time to float to the top of the holding bath.

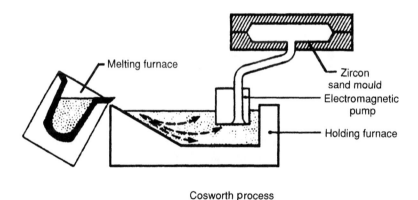

Cosworth process

Figure 11.3 *Principle of the Cosworth process. (J. Campbell,* Foundry International, *March 1992.)*

An electromagnetic pump is used to deliver clean metal from below the surface of the holding bath to the sand mould. In the original Cosworth process, productivity was low because the mould had to be kept in place on the casting machine until the casting had solidified. A later development introduced a turnover type of casting machine, allowing the casting rate to be increased by a factor of 3 or 4 to more than 1 mould per minute. The inversion of the mould also assists the feeding of the casting shrinkage, since hot feed metal ends up above the cooler casting. The expensive zircon sand is thermally reclaimed with a cycle loss claimed to be below 1%. The electromagnetic metal pump is not entirely trouble free and a breakdown, due perhaps to refractory failure, can cause serious production delays.

Other core assembly processes

Other core assembly processes use silica sand which has lower chilling power than zircon thus allowing thinner wall castings to be made. The Cosworth process is limited to casting wall thicknesses of about 4 mm, but silica sand allows castings with walls only 2.5–3 mm to be made. Although silica sand has a high thermal expansion, particularly through the alpha/beta phase transformation, Fig. 11.4, the sand moulds and cores do not reach this temperature, so thermal expansion does not seriously affect dimensions.

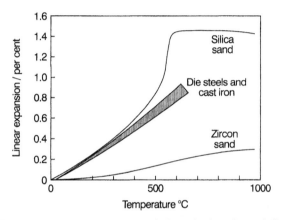

Figure 11.4   *The expansion/temperature relations for iron-based die materials, and silica and zircon sand moulding materials. (J. Campbell,* Foundry International, *March 1992)*

Gravity fill is more commonly used than a metal pump in modern core assembly casting. One patented process uses gravity filling through a down-sprue to the bottom of the mould. The moulds are purged with inert atmosphere. The core package is located from below against the bottom of the pouring spout at the bottom of the pouring vessel which prevents any contact between the metal and the atmosphere. Pouring is automatically controlled. Immediately after pouring, the moulds are rotated allowing efficient feeding with the hottest metal at the top of the mould.

The use of ceramic foam filters enables the turbulence-free filling pioneered by Cosworth to be achieved at lower cost and greater reliability. The same efficient feeding with high yield, turbulence-free filling can be achieved in a simple manner by the use of the KALPUR sleeve/filter unit, Figs 11.5a and b (see Chapter 8).

All core assembly methods use expensive chemical sand binders, usually phenolic-isocyanate cold-box resin hardened by amine gas. It is usual to reclaim the sand after use by a thermal process such as a gas-fired fluid bed

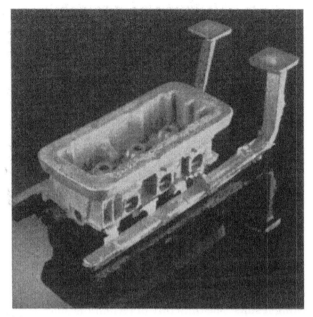

(a)

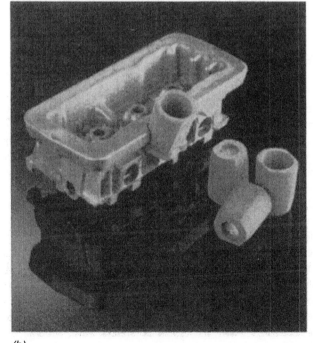

(b)

Figure 11.5 *A sand cast cylinder head: (a) Cast using a conventional running system. (b) Direct poured through a KALPUR sleeve/filter.*

reclaimer. The cost is relatively high but lower than the combined cost of using new sand and disposing of the waste.

Because of the high production cost of core assembly methods, the processes are used for complex castings such as cylinder heads and blocks where the precision and design freedom of core assembly can give significant design benefits. It is possible, for example, to integrate into a cylinder block casting a number of parts such as water pump housing, oil filter housing and brackets which would normally be cast separately then machined and bolted on. Machining and assembly is eliminated and the integrally cast unit is less likely to have leaks.

The same melting and metal treatment methods used for gravity diecasting are used (see Chapter 6).

Self-setting sand moulds

When aluminium castings are required in smaller numbers, self-setting (or sometimes gas-hardened) chemically bonded sand moulds and cores may be used. Sodium silicate, CO_2 or ester hardened, was used and gave excellent casting results. The difficulty of reclaiming silicate-based sands has led to their replacement, in many cases, by resin binders. Alkali phenolic resins such as FENOTEC and furanes such as FUROTEC can be used effectively for aluminium casting (see Chapter 12).

The Lost Foam process

Unlike any other sand casting process, no binders are used. Preforms of the parts to be cast are moulded in expanded polystyrene (EPS) using aluminium tooling. Complex shapes can be formed by gluing EPS mouldings together. The preforms are assembled into a cluster around an EPS sprue then coated with a refractory paint. The cluster is invested in dry sand in a simple moulding box and the sand compacted by vibration. Metal is poured, vaporising the EPS preform and replacing it to form the casting, Fig. 11.6. Unlike all other aluminium casting processes, the filling rate of the mould is not determined by the geometry of the running system, but by the rate at which the EPS pattern is destroyed by the liquid alloy. This in turn is greatly affected by the properties, particularly permeability, of the coating. The result is an essentially turbulence-free mode of filling, whether from the bottom, side or top. Due to the freedom from turbulence and its associated trapping of oxide films, lost foam aluminium castings can be of high metallurgical integrity. The process is therefore increasingly used for critical automotive castings such as cylinder heads and blocks, water pump housings, brackets, inlet manifolds etc. up to about 20 kg weight.

The EPS raw bead used for casting preforms is purchased from a chemical supplier. It consists of spherical beads of polystyrene of carefully graded size

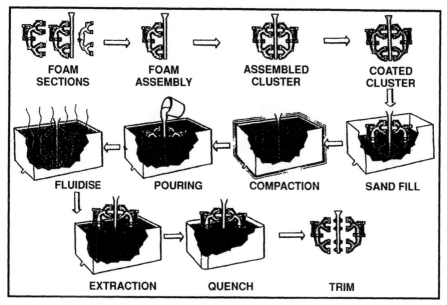

Figure 11.6 *The Lost Foam casting process.* (The Foundryman **87**, *June 1994,* p. 227.)

and type chosen specially for casting. The bead is impregnated during manufacture with a blowing agent, pentane. The bead is first pre-expanded to a precisely controlled density by steam heating. The pre-expanded bead is then moulded in a press rather like a plastic injection moulding press. The moulding tool is made of aluminium, hollow backed to have a wall thickness of around 8 mm. The pre-expanded bead is blown into the closed die, which is then steam heated causing the beads to expand further and fuse together. After fusing, the die is cooled with water sprays (often with vacuum assistance) so that the pattern is cooled sufficiently to be ejected without distortion. The moulding process takes 1–2 minutes for a cycle. The moulding machines are large, with pattern plate dimensions typically 800 × 600 mm or 1000 × 700 mm so that multiple impression dies can be used to increase the production rate.

The casting reproduces in astonishing detail the surface appearance of the EPS pattern and a great deal of effort has been put into developing special moulding techniques to minimise "bead-trace". Specialist foam pattern manufacturers such as Foseco-Morval in North America have developed moulding technology specifically for foundry applications.

Where possible, patterns are moulded in one piece using tooling techniques developed for plastic injection moulding such as metal "pull-backs" and collapsible cores, but many of the complex shapes needed to make castings cannot be moulded in one piece. Sections of patterns are glued together quickly and precisely with hot-melt adhesives using special glue-printing machines, Fig. 11.7.

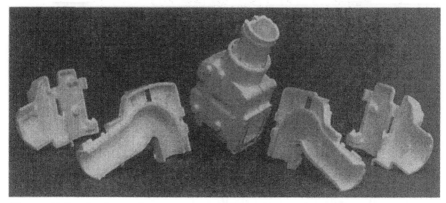

Figure 11.7   *Sections of patterns are glued together to form complex shapes.*

Some large castings are made singly, the EPS pattern being attached to a down-sprue of EPS which may be moulded hollow. Smaller castings are made in clusters, the patterns being assembled around the sprue with the ingates acting also as supports for the cluster, Fig. 11.6.

The foam pattern is covered with a refractory coating before casting. The coating must also be permeable to allow the gaseous and liquid products of thermal degradation of the pattern to escape from the mould. The permeability of the coating controls to a large extent the rate at which the mould fills. Foseco supplies the STYROMOL range of water-based coatings of carefully controlled permeability which are applied by dipping and thoroughly dried before casting.

The coated and dried pattern clusters are placed in a steel box and dry silica sand poured around. As the box is filled, it is vibrated to compact the sand and cause it to flow into the cavities of the pattern. Following much research by equipment suppliers, good vibration techniques have been developed, and patterns with complex internal form can now be reliably invested. It is not possible to persuade sand to flow uphill by vibration so patterns must be orientated in such a way that internal cavities are filled horizontally or downwards.

When liquid aluminium is cast, the heat from the metal melts and partially vaporises the EPS. The liquid and gaseous products of pyrolysis escape through the coating, the rate of escape controlling the rate at which the casting fills. The result is that the casting fills gently with very little turbulence but the time of fill is a little longer than for conventional casting. No smoke or fume is seen, the pyrolysis products condensing in the sand.

Patterns must be orientated primarily to allow complete filling with sand. Ingates fulfil the dual role of supporting the fragile pattern cluster and controlling the metal flow. When the solidified casting is removed, the sand is cooled and returned for reuse. Some contamination of the sand with styrene and other products of pyrolysis of the foam pattern occurs and a proportion of the used sand is usually thermally reclaimed to avoid a build-up of residues which could affect the flowability of the sand.

Advantages of the process

1 Low tooling cost: Though tools are expensive, their life is long, up to 500 000 cycles are possible. So for long-running, high volume parts overall tool costs are much lower than for conventional casting process. For shorter running parts the advantage is less and may even be a disadvantage.
2 Reduced finishing: There is a major advantage on most castings since finishing is restricted to removing ingates.
3 Reduced machining: For many applications, machining is greatly reduced and in some cases eliminated completely.
4 Ability to make complex castings: For suitable applications, the ability to glue patterns together to make complex parts is a major advantage.
5 Reduced environmental problems: Lost Foam is fumefree in the foundry and the sand, which contains the EPS residues, is easily reclaimed using a simple thermal process.

Disadvantages

1 The process is difficult to automate completely; cluster assembly and coating involve manual labour unless a complete casting plant is dedicated to one casting type so that specialised mechanical handling can be developed.
2 Methoding the casting is not easy and a good deal of experimentation is needed before a good casting is achieved.
3 Cast-to-size can be achieved but only after several tool modifications because the contractions of foam and casting cannot yet be accurately predicted.
4 Because of 2 and 3, long lead times are inevitable for all new castings.
5 It originally proved difficult to achieve the highest metallurgical quality in Al castings because of the need to cast at rather higher than normal temperatures. The problem has now been largely overcome through the development of thermally insulating coatings for the patterns which allow lower casting temperatures to be used.

Applications

The usual alloys used for sand and gravity casting can be cast success-fully by Lost Foam and the methods of melting and treatment are the same as those described in Chapter 6. The automotive industry is the biggest user of Lost Foam. The inlet manifold was the first successful high volume application. Cylinder heads are being made in growing numbers. Use of Lost Foam gives the designer rather more freedom to cool the working face effectively, the combustion chambers can be formed "as-cast", avoiding an expensive machining operation, and bolt holes can be cast. Lost Foam offers significant design advantages over other casting processes for cylinder blocks; features can be cast in, such as the water pump cavity, alternator bracket, oil filter mounting pad. Oil feed, drain and coolant lines can also be cast more effectively. A variety of other automotive parts are being made including water pump housings, brackets, heat exchangers, fuel pumps, brake cylinders.

Chapter 12

Sands and sand bonding systems

Silica sand

Most sand moulds and cores are based on silica sand since it is the most readily available and lowest cost moulding material. Other sands are used for special applications where higher refractoriness, higher thermal conductivity or lower thermal expansion is needed.

Properties of silica sand for foundry use

Chemical purity

SiO_2	95–96% min	The higher the silica the more refractory the sand
Loss on ignition	0.5% max	Represents organic impurities
Fe_2O_3	0.3% max	Iron oxide reduces the refractoriness
CaO	0.2% max	Raises the acid demand value
K_2O, Na_2O	0.5% max	Reduces refractoriness
Acid demand value to pH_4	6 ml max	High acid demand adversely affects acid catalysed binders

Size distribution

The size distribution of the sand affects the quality of the castings. Coarse-grained sands allow metal penetration into moulds and cores giving poor surface finish to the castings. Fine-grained sands yield better surface finish but need higher binder content and the low permeability may cause gas defects in castings. Most foundry sands fall within the following size range:

Grain fineness number	50–60 AFS ⎱	Yields good surface
Average grain size	220–250 microns ⎰	finish at low binder levels
Fines content, below 200 mesh	2% max	Allows low binder level to be used
Clay content, below 20 microns	0.5% max	Allows low binder levels
Size spread	95% on 4 or 5 screens	Gives good packing and resistance to expansion defects
Specific surface area	120–140 cm^2/g	Allows low binder levels
Dry permeability	100–150	Reduces gas defects

Grain shape

Grain shape is defined in terms of angularity and sphericity. Sand grains vary from well rounded to rounded, sub-rounded, sub-angular, angular and very angular. Within each angularity band, grains may have high, medium or low sphericity. The angularity of sand is estimated by visual examination with a low power microscope and comparing with published charts, Fig. 12.1.

The best foundry sands have grains which are rounded with medium to high sphericity giving good flowability and permeability with high strength at low binder additions. More angular and lower sphericity sands require higher binder additions, have lower packing density and poorer flowability.

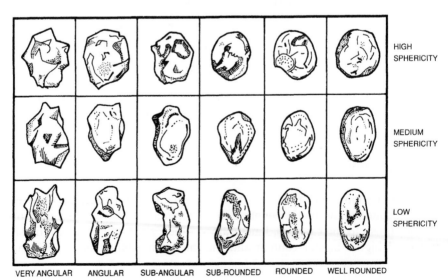

Figure 12.1 *Classification of grain shapes.*

Acid demand

The chemical composition of the sand affects the acid demand value which has an important effect on the catalyst requirements of cold-setting, acid-catalysed binders. Sands containing alkaline minerals and particularly significant amounts of seashell will absorb acid catalyst. Sands with acid demand values greater than about 6 ml require high acid catalyst levels; sands with acid demand greater than 10–15 ml are not suitable for acid catalysed binder systems.

Typical silica foundry sand properties

Chelford 60 sand (a sand commonly used in the UK as a base for green sand and for resin bonded moulds and cores)

Grain shape: rounded, medium sphericity
Bulk density, loose: 1490 kg/m^3 (93 lb/ft^3)
GF specific surface area: 140 cm^2/g
Angle of repose: 33°
Chemical analysis:

SiO$_2$	Fe$_2$O$_3$	Al$_2$O$_3$	K$_2$O	Na$_2$O	CaO	TiO$_2$	Cr$_2$O$_3$	LOI
97.1	0.11	1.60	0.73	0.15	0.10	0.06	15 ppm	0.3

Acid demand (number of ml 0.1 N HCl):

to:	pH3	pH4	pH5	pH6	pH7
ml:	2.0	1.8	1.4	1.0	0.8

Sieve grading of Chelford 60 sand:

Aperture size (μm)	BSS mesh no.	% wt. retained
1000	16	nil
700	22	0.4
500	30	2.3
355	44	10.0
250	60	25.7
210	72	23.8
150	100	28.7
105	150	7.6
75	200	1.3
–75	–200	0.2

AFS grain fineness no. 59
Base permeability: 106

Table 12.1 gives size gradings of typical foundry sands used in the UK and Germany.

Table 12.1   Typical UK and German foundry sands

Sieve size		Sand type				
		UK sands		*German sands*		
microns	*BSS no.*	*Chelford 50*	*Chelford 60*	*H32*	*H33*	*F32*
1000	16	trace	nil			
700	22	0.7	0.4			
500	30	4.5	2.3	1.0	0.5	1.0
355	44	19.8	10.0	15.0	7.5	7.0
250	60	44.6	25.7	44.0	30.0	30.0
210	72	21.6	23.8	39.0	60.0	60.0
150	100	8.2	28.7			
100	150	2.6	7.6			
75	200	nil	1.3	1.0	2.0	2.0
75	−200	nil	0.2	nil	nil	nil
AFS grain fineness no.		46	59	51	57	57
Average grain size mm		0.275	0.23	0.27	0.23	0.23

Note: Haltern 32, 33 and Frechen 32 are commonly used, high quality German sands.
German sieve gradings are based on ISO sieves.
The German sands have rounder grains and are distributed on fewer sieves than UK sands, they require significantly less binder to achieve the required core strength.

Safe handling of silica sand

Fine silica sand (below 5 microns) can give rise to respiratory troubles. Modern foundry sands are washed to remove the dangerous size fractions and do not present a hazard as delivered. It must be recognised, however, that certain foundry operations such as shot blasting, grinding of sand covered castings or sand reclamation can degrade the sand grains, producing a fine quartz dust having particle size in the harmful range below 5 microns. Operators must be protected by the use of adequate ventilation and the wearing of suitable face masks.

Segregation of sand

Segregation, causing variation of grain size, can occur during sand transport or storage and can give rise to problems in the foundry. The greatest likelihood of segregation is within storage hoppers, but the use of correctly designed hoppers will alleviate the problem:

1 Hoppers should have minimum cross-sectional area compared to height,
2 The included angle of the discharge cone should be steep, 60–75°,
3 The discharge aperture should be as large as possible,

Measurement of sand properties

Acid demand value

Acid demand is the number of ml of 0.1 M HCl required to neutralise the alkali content of 50 g of sand:

Weigh 50 g of dry sand into a 250 ml beaker
Add 50 ml of distilled water
Add 50 ml of standard 0.1 M hydrochloric acid by pipette
Stir for 5 minutes
Allow to stand for 1 hour
Titrate with a standard solution of 0.1 M sodium hydroxide to pH values of 3, 4, 5, 6 and 7
Subtract the titration values from the original volume of HCl (50 ml) to obtain the acid demand value

Grain size

See Chapter 1 for the method of measuring average grain size and AFS grain fineness number.

Thermal characteristics of silica sand

Silica sand has a number of disadvantages as a moulding or coremaking material:

It has a high thermal expansion rate (Fig. 12.2) which can cause expansion defects in castings, such as finning or veining and scabbing
It has a relatively low refractoriness, Table 12.2, which can cause sand burn-on, particularly with steel or heavy section iron castings
It is chemically reactive to certain alloys for example ferrous alloys containing manganese. The oxides of Mn and Fe react with silica to form low melting point silicates, leading to serious sand burn-on defects

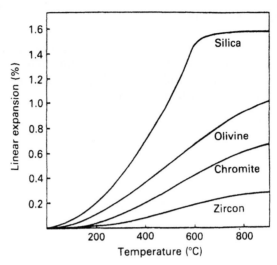

Figure 12.2   *Thermal expansion characteristics of zircon, chromite and olivine sands compared with silica sand.*

Table 12.2   Sintering points of silica sand

Sand	Sintering point (°C)
High purity silica sand, >99% quartz	1450
Medium purity silica sand, 96% quartz	1250
Sea sand (high shell content)	1200
Natural clay bonded sand	1050–1150

For certain types of casting, it may be necessary to use a non-silica sand, even though all other sands are more expensive than silica.

Non-silica sands (Table 12.3)

Zircon, ZrSiO$_4$

Zircon sand has a high specific gravity (4.6) and high thermal conductivity which together cause castings to cool faster than silica sand. The chilling effect of zircon sand can be used to produce favourable thermal gradients that promote directional solidification giving sounder castings.

Table 12.3 Properties of non-silica sands (compared with silica)

Property	Silica	Zircon	Chromite	Olivine
AFS grain size no.	60	102	74	65
Grain shape	rounded	rounded	angular	angular
Specific gravity	2.65	4.66	4.52	3.3
Bulk density (kg/m³)	1490	2770	2670	1700
(lb/ft³)	93	173	167	106
Thermal expansion 20–1200°C	1.9% non linear	0.45%	0.6%	1.1%
Application	general	refractoriness chill	resistance to penetration chill	Mn steel

The thermal expansion coefficient of zircon is very low (Fig. 12.2) so that expansion defects can be eliminated. Zircon has higher refractoriness than silica; moreover it does not react with iron oxide, so sand burn-on defects can be avoided. Zircon sand generally has a fine grading, with AFS number between 140 and 65 (average grain size 115–230 microns); the most frequently used grade is around AFS 100.

Zircon is probably the most widely used of the non-silica sands. It is used with chemical binders for high quality steel castings and for critical iron castings such as hydraulic spool valves which contain complex cores, almost totally enclosed by metal, making core removal after casting difficult. Zircon has low acid demand value and can be used with all chemical binder systems. The Cosworth casting process uses the low thermal expansion of zircon sand cores and moulds to cast dimensionally accurate castings. The high cost of zircon sand makes reclamation necessary and thermal reclamation of resin bonded moulds and cores is frequently practised.

Zircon sands contain low levels of naturally occurring radioactive materials, such as uranium and thorium. Any employer who undertakes work with zircon mineral products is required, by law, to restrict exposure of workers to such naturally occurring contaminants so far as reasonably practical. The primary requirement is to prevent the inhalation of zircon dust. Suitable precautions are set out in the UK in HSE Guidance Note EH55: Dust – general principles of protection. Guidance on the possible hazards associated with the use, handling, processing, storage, transport or waste disposal of such naturally occurring radioactive materials and the control measures that are recommended to minimise exposure should be obtained from the supplier.

Chromite, $FeCr_2O_4$

The high specific gravity (4.5) and high thermal conductivity of chromite provide a pronounced chilling effect when it is used for moulds and cores. Thermal expansion is low so expansion defects are unlikely to occur. Chromite sand has a glossy black appearance; it has greater resistance to metal penetration than zircon in spite of its generally coarser grading (typically AFS 70). It has somewhat higher acid demand than other sands which entails greater additions of acid catalyst when furane resin is used. Apart from this the sand is compatible with all the usual binder systems. Chromite is generally used for steel casting to provide chilling. It is difficult to reclaim chromite sand since, if it becomes contaminated with silica, its refractoriness is seriously reduced.

Olivine, Mg_2SiO_4

Olivine sand is used mainly for the production of austenitic manganese steel castings (which react with silica and other sands to give serious burn-on defects). It has also been used to avoid the health hazards possible with silica sand. Olivine has a very high acid demand and is not suitable for use with acid-catalysed binders such as furane resins; it tends to accelerate the curing of phenolic urethane binders. Being a crushed rock, it is highly angular and consequently requires high binder additions. Thermal expansion is regular and quite low.

Green sand

The earliest method of bonding sand grains to form a sand mould was to use clay and water as a binder. The moulds could be used in the "green" or undried state (hence the term green sand moulding) or they could be baked in a low temperature oven to dry and strengthen them to allow heavy castings to be made. Nowadays, dried, clay bonded sand is little used, having been replaced by chemically bonded sand, but green sand is still the most widely used moulding medium, particularly for iron castings.

Originally, naturally occurring clay sand mixtures were used, containing 10% or more of clay. It was found that by adding coal dust, the ease of stripping iron castings from the mould and the surface finish of the castings could be greatly improved. The heating of the coal dust by the liquid iron causes the formation of a type of carbon called lustrous carbon which is not wetted by the liquid iron, so the cast surface is improved. Green sand used for aluminium casting does not require coal dust additions.

Clay bonded moulding sand can be used over and over again by adding water to replace that which is lost during casting, and remilling the sand. However, clay which is heated to a high temperature becomes "dead", that is, it loses its bonding power. The coal dust is partly turned to ash by heat, so new clay, coal dust and water must be added and the sand remilled to restore its bonding properties. As the sand is reused, dead clay and coal ash build up in the sand, reducing its permeability to gases so that eventually water vapour and other mould gases are unable to escape from the mould and defective castings are produced.

Natural sands are little used nowadays (except for some aluminium castings) and most foundry green sands are "synthetic" mixtures based on washed silica sand to which controlled additions of special moulding clays (bentonites) and low ash coal dust are made. Alternatively, special blended additives such as BENTOKOL, which combine the clay binder with lustrous carbon formers, can be used. The moulding sand becomes a "sand system" which is continuously recycled, with suitable additions and withdrawals made as required. The control of system sand to achieve constant moulding properties has become an important technology of its own for it determines the quality of the castings produced.

Green sand additives

Base sand

Silica sand of AFS grain size 50–60 is usually used. The particle size distribution is important, a sand spread over 3 to 5 consecutive sieve sizes with more than 10% on each sieve gives the best results (see Table 12.1). Rounded or sub-angular sands are the best since the more round the grains, the better the flowability and permeability of the sand. The base sand should have the same size grading as the core sand used in the foundry, so that burnt core sand entering the system does not alter the overall size grading.

Clay

The best bonding clays are bentonites which can either have a calcium or a sodium base. Sodium bentonites occur naturally in the USA as Wyoming or Western bentonite and also in other countries particularly in the Mediterranean area. Green sand produced with this clay has medium green strength and high dry strength which increases the resistance to erosion of metal but can give problems at shakeout. Sodium bentonites are commonly used for steel casting production.

Calcium bentonites are more widely distributed, they produce green sands with rather high green strength but low dry strength so they have low erosion resistance and are prone to scabbing and other expansion defects. Calcium bentonites can be converted to sodium bentonite by adding soda ash, the calcium base is replaced by sodium and the clay then has properties approaching those of natural sodium bentonite. Such clays are known as activated clays.

Blended or mixed bentonites are commercial blends of sodium bentonite with calcium bentonite or a sodium-activated bentonite. Compositions can be varied to suit particular applications. Most iron and non-ferrous foundries use blended bentonites.

Clays absorb water from the atmosphere so they should be stored in dry conditions, they do not deteriorate on storage, even for long periods.

Coal dust

Used mainly in iron foundries with some being used in non-ferrous foundries. The formation of "lustrous carbon" from the thermal degradation of the volatiles given off during casting improves casting surface finish and strip. Good coal dust has the following properties:

Volatile content:	33–36%
Ash:	less than 5%
Fixed carbon:	50–54%
Sulphur:	less than 1%
Chlorine:	less than 0.03%
Size:	75 or 100 grade (AFS grain fineness no.)

It is important that the size grading of the coal dust used should not be too fine. Coal dust increases the moisture requirement of the sand and the finer it is, the more moisture is needed, which may have harmful effects on the castings. Fine coal dust will also reduce the overall permeability of the sand. Coal dust levels in green sand vary from 2 or 3% for small iron castings to 7 or 8% for heavy section castings. Too much coal dust can give rise to gas holes in the castings or misruns.

Coal dust must be stored dry to prevent the risk of fire. Damp coal dust can ignite spontaneously. Large stocks are undesirable and should be rotated first in, first out.

Coal dust replacements

Consist of blends of highly volatile, high lustrous carbon materials blended with clays. They are generally more environmentally acceptable than coal dust, producing less fume during casting. They can be designed for particular applications such as high pressure moulding.

BENTOKOL

BENTOKOL additives are blends of natural clays to which have been added specially selected essential volatiles to provide both bond and volatile content for iron foundry green sand. They are in powder form, virtually dust free with flow properties and bulk densities similar to clay, they are supplied in bulk. Use of BENTOKOL often means that a "one shot" sand addition is sufficient and always improves working conditions. Fume after casting is reduced and the foundry is cleaner and pleasanter while casting finish is maintained or enhanced.

Many of the BENTOKOL formulations contain a blend of natural sodium bentonite which has high burn-out temperature and results in lower concentrations of dead clay. This improves permeability, reduces moisture requirements and gives better all-round sand properties over a longer life cycle. Sodium bentonites take longer than other clays to develop their bond. To offset this BENTOKOL formulations may include dispersion agents so that the clay is rapidly spread and the bond quickly developed.

BENTOKOL is compatible with sand systems based on clay and coal dust. Certain parameters of the sand system may change with the establishment of BENTOKOL; these may be a slight reduction of volatiles, LOI and clay fraction and the optimum moisture content will fall by up to 0.5%.

Starches, cereals and dextrines

Used mainly in steel foundries to increase strength and toughness of the green sand. Starches can help to reduce expansion defects, since as they burn out, they allow the sand grains to expand without deforming the mould. Cereals increase green strength, dry strength and toughness but can reduce flowablity. Dextrines improve flowability and moisture retention, preventing moulds from drying out and edges becoming friable.

Water

Water is needed to develop the clay bond but it can cause casting defects. Where there is strong, localised heating, e.g. in the vicinity of ingates or on horizontal mould surfaces exposed to radiant heat from the metal, moisture is driven back from the mould surface, condensing in a wet, weak underlying layer that can easily fracture to produce expansion defects in castings such as scabs, rat-tails and buckles.

MIXAD additive

MIXAD 61 additive has been developed to eliminate sand expansion defects. By improving the wet strength of the condensed water layer, MIXAD 61 eliminates the associated defects. MIXAD 61 is added early in the milling cycle, usually 0.5–1.0% is used in the facing sand. If MIXAD 61 is used in a unit sand system, an addition of 0.05–0.10% is usually sufficient. MIXAD 61 is used principally in sands for iron and copper-based alloys.

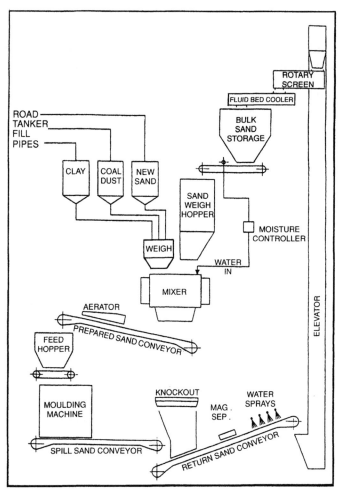

Figure 12.3 *Flow diagram for a typical green sand plant. (Courtesy Foundry and Technical Liaison Ltd.)*

The green sand system

The "sand system" in green sand foundries is illustrated in Fig. 12.3. It comprises the following items of equipment:

The bulk sand storage hopper

Returned knockout sand is stored in a hopper which ideally has capacity for about 4 hours of usage. Returned sand is very variable in properties, since some will have been heated very strongly by the hot metal while sand which came from the edges of the mould will have been heated very little. Some mixing of the sand occurs in the storage hopper. It is better to have several smaller volume hoppers rather than one large unit since mixing of the return sand is improved. Hoppers should be designed in such a way that sand flows evenly through without sticking to the walls.

The sand mill

At the mill, the additions of new sand, clay, coal dust and water are made to the returned sand. Sand mills may be continuous or batch but batch mills are nowadays preferred because better sand control is possible. Several designs of mill are available, their purpose is to mix the sand and spread the moistened clay over the surface of the sand grains in order to develop the bond. Older sand mills used heavy vertically set mulling wheels and plough blades (Fig. 12.4). A certain optimum milling time is necessary and this has to be determined for each green sand system and adhered to carefully. Short milling times underuse the additives and produce sand having low and variable green strength. Modern sand mills use intensive mixers to develop the clay bond quickly and efficiently. Whatever type of mill is used, regular maintenance is essential to ensure that the milling efficiency does not change.

Additions of new sand, clay and coal dust (or coal dust replacement) are made at the mill. Batch mills use weighed additions while continuous mills use calibrated feeders to make continuous additions.

Water is added, often via an automatic moisture controller which measures the moisture content of the sand as it feeds into the mixer batch hopper prior to mixing then calculates the exact amount of water to be added to the sand batch discharged into the mixer to bring the moisture to the required level.

Prepared sand is usually passed by an aerator which "fluffs up" the sand before feeding the moulding machines.

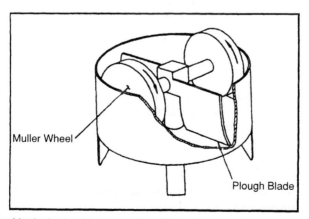

Figure 12.4 *Vertical wheel batch muller. (Sixth Report of Institute Working Group T30, Mould and Core Production, Foundryman, Feb. 1996.)*

Buffer hopper

A buffer hopper is usually provided before the moulding machines, so that a moulding machine stoppage does not necessitate running freshly milled sand straight back to the main hopper.

The moulding machine(s)

These can be jolt squeeze, high pressure, impact moulding machines etc. Each has its advantages and each requires special moulding sand properties.

Casting

The moulds are cast and allowed to cool for a suitable time, often 30–40 minutes, before knocking out the castings. The cooling time not only allows the castings to solidify completely but also reduces internal stresses within the casting caused by differential cooling of sections of varying thickness.

Shake out

The shakeout separates the castings from the sand. It may be a vibrating grid or a rotating drum, the latter also having a cooling function since water is sprayed onto the sand in the drum. The shakeout is provided with copious air extraction, to prevent dust from entering the foundry. This also removes some fines from the sand and so plays an important part in the control of the sand system.

Return sand conveyor

Sand from the shakeout contains lumps of partially burnt cores, very hot burnt core sand, hot moulding sand from near the casting and cooler moulding sand from the edges of moulds. A magnetic separator removes ferrous metallic particles and water sprays are used to effect some preliminary cooling before elevation, since very hot sand can damage the elevator.

Screen

The return sand is elevated to a screen where residual core lumps and other unwanted material are removed before the sand is returned to the hopper. Non-magnetic metallics are also removed by the screen.

Sand cooling

Hot sand causes problems of excessive moisture loss and condensation of moisture on patterns and cores, so the returned sand must be cooled. The only effective way of cooling green sand is by evaporation of water. Evaporation of 1% water cools the sand by about 25°C, so water is sprayed onto the sand and air drawn over the sand to evaporate it, either in the rotary drum shakeout or at the screen or in a specially designed fluidised bed cooler. Returned green sand cannot be fluidised by air alone. Sand is delivered onto a screen which is vibrated to transport the sand along it. Water is added by sprays and evaporated by blowing air through the perforated screen.

Sand removal or addition

The sand system must have a facility for removal of excess sand, since foundries making highly cored castings build up the sand level in the system through the introduction of burnt core sand. Regular addition of new sand is needed to maintain control of green sand properties.

Green sand properties

Green sand typically has the following properties:

	jolt/squeeze machines	high pressure (DISA etc.)
water	3–4%	2.5–3.2%
green strength	70–100 kPa	150–200 kPa
	10–15 psi	22–30 psi
compactability	45–52%	38–40%
permeability	80–110	80–100
live clay	5.0–5.5%	6.0–10.0%
volatiles	2.5%	2.0%
LOI	7.0–7.5%	6.0%

A typical grading of a sand suitable for a green sand system is:

Sieve mesh size (μm)	%
710	0.14
500	1.42
355	5.44
250	22.88
212	19.64
150	26.74
106	9.70
75	2.84
pan	1.10

Green sand for aluminium casting does not need volatile additions and a finer grain size base sand may be used to improve casting surface finish.

Control of green sand systems

When metal is poured into green sand moulds, the heat from the metal drives off some water from the sand and clay, burns some of the coal dust to coke and ash, and burns a proportion of the clay so that its bonding properties are destroyed. Before the sand can be used again its properties must be restored by removing burnt clay, coal dust and ash, and milling in new clay, coal dust and water. The following additions are usually necessary at the sand mill in order to maintain the required moulding properties of the sand:

0.3–0.5% of new clay
0.3–0.5% of coal dust
1.5–2.5% of water

In addition, it is desirable to add a proportion (up to 10%) of new silica sand to the system and to dispose of a corresponding amount of used system sand. If highly cored work is made, the amount of new sand addition can possibly be reduced since well-burnt core sand may serve as new sand, but care must be taken since certain core binder residues (particularly from phenolic-isocyanate binders) can harm the green sand properties.

The precise amount of the additions that must be made depends on many factors, including:

The weight and type of castings being made, which affects the burn-out
The amount of core sand entering the system
Whether hold-ups on the moulding line have allowed freshly milled sand to return directly to the sand hopper

Good sand control depends on carefully monitored experience of the particular sand system, since every system is different. The normal additions of clay, coal dust or BENTOKOL at each cycle represent only about 10% of the total active clay or coal present in the system. It is not possible to change the total clay or coal dust content quickly since any change in the addition rate takes about 20 cycles to work its way fully into the system. For example, an addition of 0.3% clay is usually sufficient to maintain the total clay level at 3.0%. If the clay addition is increased to 0.4%, the total clay content after one cycle will only rise to about 3.1% and it will take 20 cycles (about 1 week) for the full effect of the change of addition to be felt and for the clay level to rise to around 4.0%.

The operator must find by experience what additions are needed during normal operation of the foundry to maintain the required physical properties of the sand; changes will normally only be necessary if some change of practice has occurred, such as using heavier castings, which will cause more burn-out of clay and coal. A change to more highly cored castings is likely to result in more burnt core sand entering the system, which will require greater additions to compensate.

Because sand systems have such built-in inertia, or resistance to change, the only way that quick changes can be made to the moulding properties of the sand is by increasing or reducing the addition of water to the sand mill. This has an immediate effect on sand properties and as long as the other constituents of the sand are in control, the sand's mouldability can quickly be corrected.

Sand testing

Regular testing of the properties of the sand is essential. One or two sand tests do not truly indicate the condition of the whole sand system, since a sand sample weighs only about 1 kg and cannot represent the whole 200 tonnes or so of the system. At least 5 samples should be taken per shift and measured for moisture, green strength, compactability and permeability. LOI and volatiles should be measured once per day. Active clay, twice per week.

Records of additions should also be kept:

Weight of clay, coal dust and new sand added each day
Number of moulds made
Weight of metal poured
Weight of used sand removed from the system each day

The 3 ram test method developed by the AFS in the 1920s is still widely used to measure green strength and compactability although it is not really suitable for sand used in modern high pressure moulding machines. The AFS ramming test is being replaced by tests on 50 mm diameter specimens produced by squeeze machines which can reproduce the pressures actually developed on the mould by the machine that the foundry uses.

Control graphs

Individual figures mean rather little, but daily average sand properties should be plotted together with weekly figures of active clay and average additions of clay, coal dust and new sand. After a few weeks of plotting the data, it will be possible to draw control lines. Variation within the lines is permissible but if results appear outside the control lines, then action must be taken such as increasing or reducing clay or coal or new sand. If action is taken, it must be remembered that it will be 1 or 2 days before the full effect of the action will be seen.

The sand properties, moisture, green strength, compactability and permeability are of first importance and must be maintained. The remaining graphs are of secondary importance, but provide useful information which may allow problems to be anticipated.

It is also necessary to keep a log book for the sand system listing information such as:

> When the sand mill was maintained
> When new sand was added and old sand removed
> Changes in foundry practice, such as change of core binder, or change in type of casting made

Parting agents

Unless a water-repellent material is present on the pattern surface to prevent wetting, green sand will tend to stick to the pattern when rammed. This will cause damage to the mould when the pattern is withdrawn. To prevent this, it is usual to spray the pattern with a liquid parting agent such as SEPAROL. The frequency of application depends on the type and quality of the pattern, but every 10th or 12th mould is typical.

Special moulding materials, LUTRON

LUTRON moulding sand is a waterless sand containing mineral oils designed for use with fine sands to produce a superfine finish in castings such as name plates, plaques, medallions and other art castings made in aluminium or copper-based alloys.

The ready-mixed LUTRON moulding sand is based on a very fine sand (AFS no. 160). The sand has excellent flowability due to the lubricating properties of the bond. It can be used alone or as a facing sand backed with normal green sand or other type of sand. If used as a unit sand, the mix can be reconstituted after casting by adding a compensating amount of LUTRON binder and milling.

The LUTRON binder can be used with any suitable, fine, dry sand; additions of 10–12% are normally needed.

Chapter 13

Resin bonded sand

Chemically bonded sand

A wide variety of chemical binders is available for making sand moulds and cores. They are mostly based either on organic resins or sodium silicate, although there are other inorganic binders such as cement, which was the earliest of the chemical binders to be used, ethyl silicate, which is used in the Shaw process and for investment casting, and silica sol, which is also used for investment casting.

The binders can be used in two ways:

> As self-hardening mixtures; sand, binder and a hardening chemical are mixed together; the binder and hardener start to react immediately, but sufficiently slowly to allow the sand to be formed into a mould or core which continues to harden further until strong enough to allow casting. The method is usually used for large moulds for jobbing work, although series production is also possible,
> With triggered hardening; sand and binder are mixed and blown or rammed into a core box. Little or no hardening reaction occurs until triggered by applying heat or a catalyst gas. Hardening then takes place in seconds. The process is used for mass production of cores and in some cases for moulds for smaller castings.

Self-hardening process (also known as self-set, no-bake or cold-setting process)

Clean, dry sand is mixed with binder and catalyst, usually in a continuous mixer. The mixed sand is vibrated or hand-rammed around the pattern or into a core box; binder and catalyst react, hardening the sand. When the mould or core has reached handleable strength (the strip time), it is removed from the pattern or core box and continues to harden until the chemical reaction is complete.

Since the binder and catalyst start to react as soon as they are mixed, the mixed sand has a limited "work time" or "bench life" during which

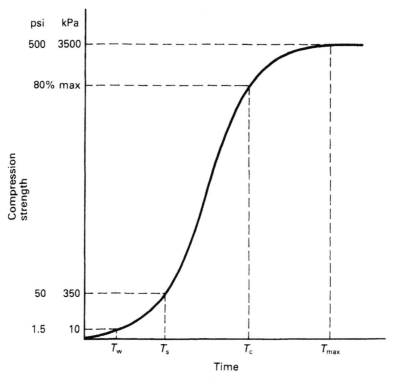

Figure 13.1   *Typical hardening curve for self-hardening sand:*

T_w = *work time*

T_s = *strip time*

T_c = *casting time*

T_{max} = *time to achieve maximum strength.*

the mould or core must be formed (Fig. 13.1). If the work time is exceeded, the final strength of the mould will be reduced. Work time is typically about one third of the "strip time" and can be adjusted by controlling the type of catalyst and its addition rate. The work time and strip time must be chosen to suit the type and size of the moulds and cores being made, the capacity of the sand mixer and the time allowable before the patterns are to be reused. With some binder systems the reaction rate is low at first, then speeds up so that the work time/strip time ratio is high. This is advantageous, particularly for fast-setting systems, since it allows more time to form the mould or core.

Stripping is usually possible when the sand has reached a compression strength of around 350 kPa (50 psi) but the actual figure used in practice depends on the type of binder system used, the tendency of the binder to sag before it is fully hardened, the quality of the pattern equipment and the complexity of the moulds and cores being made.

It is advisable to strip patterns as soon as it is practical, since some binder chemicals attack core box materials and paints after prolonged contact. The properties of chemical binders can be expressed in terms of:

> *Work time (bench life):* which can be conveniently defined as the time after mixing during which the sand mixture has a compressive strength less than 10 kPa, at this stage it is fully flowable and can be compacted easily,
>
> *Strip time:* which can be defined as the time after mixing at which a compressive strength of 350 kPa is reached, at this value most moulds and cores can be stripped without damage or risk of distortion,
>
> *Maximum strength:* the compressive strength developed in a fully hardened mixture, figures of 3000–5000 kPa are often achieved

It is not necessary to wait until the maximum strength has been achieved before moulds can be cast, the time to allow depends on the particular castings being made; usually casting can take place when 80% of the maximum strength has been reached.

Testing chemically bonded, self-hardening sands

Units

Compressive strength values may be reported in:

SI units	$kPa = kN/m^2$
cgs units	kgf/cm^2
Imperial units	$psi = lbf/in^2$

Conversion factors:

$$100\,kPa\,(kN/m^2) = 1.0197\,kgf/cm^2$$
$$= 14.5038\,psi\,(lbf/in^2)$$

$$1\,kgf/cm^2 = 98.0665\,kPa$$
$$= 14.22\,psi\,(lbf/in^2)$$

$$1\,psi\,(lbf/in^2) = 6.895\,kPa\,(kN/m^2)$$
$$= 0.07032\,kgf/cm^2$$

Conversion table

kPa (kN/m²)	kgf/cm²	psi (lbf/in²)
10	0.10	1.5
50	0.51	7.3
100	1.02	14.5
200	2.04	29.0
300	3.06	43.5
400	4.08	58.0
500	5.10	72.5
600	6.12	87.0
700	7.14	101.5
800	8.16	116.0
900	9.18	130.5
1000	10.20	145.0
2000	20.39	290.1
3000	30.59	435.1
4000	40.79	580.1
5000	50.99	725.2

The curing properties (work time, strip time and maximum strength) are measured by compression tests using 50 mm diameter specimen tubes with end cups, or AFS 2 in diameter tubes, with a standard rammer. Sand is mixed in a food mixer or small core sand mixer; catalyst being added first and mixed, then the resin is added and mixed.

Measurement of "work time" or "bench life"

Mix the sand as above, when mixing is complete, start a stopwatch and discharge the sand into a plastic bucket and seal the lid.

After 5 minutes, prepare a standard compression test piece and immediately measure the compressive strength.

At further 5 minute intervals, again determine the compressive strength, stirring the mixed sand in the bucket before sampling it.

Plot a graph of time v. strength and record the time at which the compressive strength reaches 10 kPa (0.1 kgf/cm², 1.5 psi); this is the worktime or bench life.

The sand temperature should also be recorded.

For fast-setting mixtures, the strength should be measured at shorter intervals, say every 1 or 2 minutes.

Measurement of strip time

Prepare the sand mixture as before.

When mixing is complete, start a stopwatch.

Prepare 6–10 compression test pieces within 5 minutes of completion of mixing the sand.

Cover each specimen with a waxed paper cup to prevent drying.

Determine the compressive strength of each specimen at suitable intervals, say every 5 minutes.

Plot strength against time.

Record the time at which the strength reaches 350 kPa (3.6 kgf/cm^2, 50 psi), this is the "strip time".

The sand temperature should also be recorded.

Measurement of maximum strength

Prepare the sand mixture as before.

Record the time on completion of mixing.

Prepare 6–10 specimens as quickly as possible covering each with a waxed cup.

Determine the strength at suitable intervals, say 1, 2, 4, 6, 12, 24 hours.

Plot the results on a graph and read the maximum strength.

The sand temperature should be held constant if possible during the test.

While compressive strength is the easiest property of self-hardening sand to measure, transverse strength or tensile strength being used more frequently nowadays, particularly for the measurement of maximum strength.

Mixers

Self-hardening sand is usually prepared in a continuous mixer, which consists of a trough or tube containing a mixing screw. Dry sand is metered into the trough at one end through an adjustable sand gate. Liquid catalyst and binder are pumped from storage tanks or drums by metering pumps and introduced through nozzles into the mixing trough; the catalyst nozzle first then binder (so that the binder is not exposed to a high concentration of catalyst).

Calibration of mixers

Regular calibration is essential to ensure consistent mould and core quality and the efficient use of expensive binders. Sand flow and chemical flow rates should be checked at least once per week, and calibration data recorded in a book for reference:

Sand: Switch off the binder and catalyst pumps and empty sand from the trough. Weigh a suitable sand container, e.g. a plastic bin holding about 50 kg. Run the mixer with sand alone, running the sand to waste until a steady flow is achieved. Move the mixer head over the weighed container and start a stopwatch. After a suitable time, at least 20 seconds, move the mixer head back to the waste bin and stop the watch. Calculate the flow in kg/min. Repeat three times and average. Adjust the sand gate to give the required flow and repeat the calibration,

Binders: Switch off the sand flow and the pumps except the one to be measured. Disconnect the binder feed pipe at the inlet to the trough, ensuring that the pipe is full. Using a clean container, preferably a polythene measuring jug, weigh the binder throughput for a given time (minimum 20 seconds). Repeat for different settings of the pump speed regulator. Draw a graph of pump setting against flow in kg/min.

Repeat for each binder or catalyst, taking care to use separate clean containers for each liquid. Do not mix binder and catalyst together, since they may react violently. Always assume that binders and catalysts are hazardous, wear gloves, goggles and protective clothing.

When measuring liquid flow rate, the pipe outlet should be at the same height as the inlet nozzle of the mixer trough, so that the pump is working against the same pressure head as in normal operation.

Mixers should be cleaned regularly. The use of STRIPCOTE AL applied to the mixer blades, reduces sand build-up.

Sand quality

In all self-hardening processes, the sand quality determines the amount of binder needed to achieve good strength. To reduce additions and therefore cost, use high quality sand having:

AFS 45–60 (average grain size 250–300 microns)
Low acid demand value, less than 6 ml for acid-catalysed systems
Rounded grains for low binder additions and flowability
Low fines for low binder additions
Size distribution, spread over 3–5 sieves for good packing, low metal penetration and good casting surface

Pattern equipment

Wooden patterns and core boxes are frequently used for short-run work. Epoxy or other resin patterns are common and metal equipment, usually aluminium, may be used for longer running work. The chemical binders used may be acid or alkaline or may contain organic solvents which can attack the patterns or paints. STRIPCOTE AL aluminium-pigmented suspension release agent or silicone wax polishes are usually applied to patterns and core boxes to improve the strip of the mould or core. Care must be taken to avoid damage to the working surfaces of patterns and regular cleaning is advisable to prevent sand sticking.

Curing temperature

The optimum curing temperature for most binder systems is 20–25°C but temperatures between 15 and 30°C are usually workable. Low temperatures retard the curing reaction and cause stripping problems, particularly if metal pattern equipment is used. High sand temperatures cause reduction of work time and poor sand flowability and also increase the problem of fumes from the mixed sand. If sand temperatures regularly fall below 15°C, the use of a sand heater should be considered.

Design of moulds using self-hardening sand

Moulds may be made in flasks or flaskless. Use of a steel flask is common for large castings of one tonne or more, since it increases the security of casting. For smaller castings, below one tonne, flaskless moulds are common. Typical mould designs are illustrated in Fig. 13.2. The special features of self-hardening sand moulds are:

Large draft angle (3–5°) on mould walls for easy stripping
Incorporation of a method of handling moulds for roll-over and closing
Means of location of cope and drag moulds to avoid mismatch
Reinforcement of large moulds with steel bars or frames
Clamping devices to restrain the metallostatic casting forces
Use of a separate pouring bush to reduce the sand usage
Mould vents to allow gas release
Sealing the mould halves to prevent metal breakout
Weighting of moulds if clamps are not used
Use of minimum sand to metal ratio to reduce sand usage, 3 or 4 to 1 is typical for ferrous castings

Foundry layout

With self-hardening sand, moulds and cores are often made using the same binder system, so that one mixer and production line can be used. A typical layout using a stationary continuous mixer is shown in Fig. 13.3. The

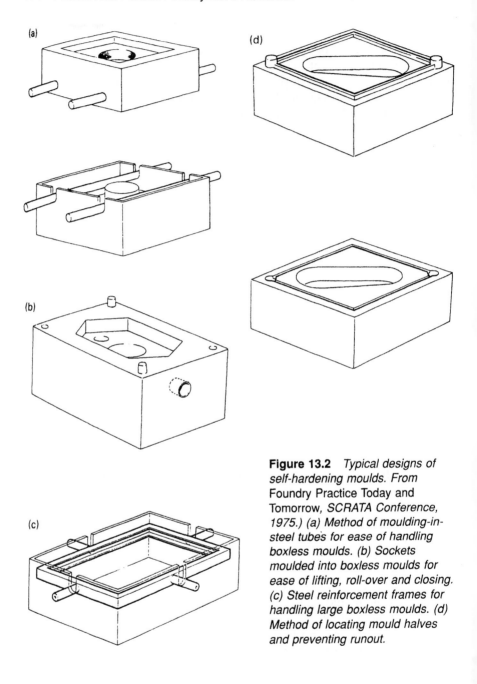

Figure 13.2   *Typical designs of self-hardening moulds. From Foundry Practice Today and Tomorrow, SCRATA Conference, 1975.) (a) Method of moulding-in-steel tubes for ease of handling boxless moulds. (b) Sockets moulded into boxless moulds for ease of lifting, roll-over and closing. (c) Steel reinforcement frames for handling large boxless moulds. (d) Method of locating mould halves and preventing runout.*

moulds may or may not be in flasks. Patterns and core boxes circulate on a simple roller track around the mixer. The length of the track is made sufficient to allow the required setting time, then moulds and cores are stripped and the patterns returned for reuse.

For very large moulds, a mobile mixer may be used.

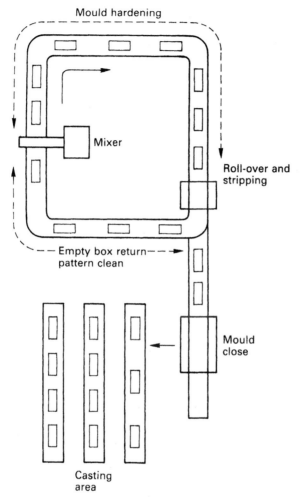

Figure 13.3 *Foundry layout for self-hardening sand moulds.*

Sand reclamation

The high cost of new silica sand and the growing cost of disposal of used foundry sand make the reclamation and reuse of self-hardening sands a matter of increasing importance. Reclamation of sand is easiest when only one type of chemical binder is used. If more than one binder is used, care must be taken to ensure that the binder systems are compatible. Two types of reclamation are commonly used, mechanical attrition and thermal.

Wet reclamation has been used for silicate bonded sand. The sand is crushed to grain size, water washed using mechanical agitation to wash off the silicate residues, then dried. The process further requires expensive water treatment to permit safe disposal of the wash water so its use is not common.

The difficulty and cost of disposing safely of used chemically bonded sand has led to the growing use of a combination of mechanical and thermal treatment. Mechanical attrition is used to remove most of the spent binder. Depending on the binder system used, 60–80% of the mechanically reclaimed sand can be rebonded satisfactorily for moulding, with the addition of clean sand. The remaining 20–40% of the mechanically treated sand may then be thermally treated to remove the residual organic binder, restoring the sand to a clean condition. This secondarily treated sand can be used to replace new sand. In some cases, all the used sand is thermally treated.

Mechanical attrition

This is the most commonly practised method because it has the lowest cost. The steps in the process are:

> Lump breaking; large sand lumps must be reduced in size to allow the removal of metal etc.
> Separation of metal from the sand by magnet or screen.
> Disintegration of the sand lumps to grain size and mechanical scrubbing to remove as much binder as possible, while avoiding breakage of grains.
> Air classification to remove dust, fines and binder residue.
> Cooling the sand to usable temperature.
> Addition of new sand to make up losses and maintain the quality of the reclaimed sand.

Reclamation by attrition relies on the fact that the heat of the casting burns or chars the resin binder close to the metal. Even at some distance from the metal, the sand temperature rises enough to embrittle the resin bond. Crushing the sand to grain size followed by mechanical scrubbing then removes much of the embrittled or partially burnt binder. The more strongly the sand has been heated, the more effectively is the sand reclaimed.

Mechanical attrition does not remove all the residual binder from the sand, so that continued reuse of reclaimed sand results in residual binder levels increasing until a steady state is reached which is determined by:

> the amount of burnout which occurs during casting and cooling
> the effectiveness of the reclamation equipment
> the percentage of new sand added
> the type of binder used

The equilibrium level of residue left on the sand is approximately expressed as:

$$P = \frac{TB}{1 - TR}$$

P is the maximum percentage of resin that builds up in the sand (the LOI of the reclaimed sand)
B is the binder addition %
T is the fraction of binder remaining after reclamation
R is the fraction of sand reused

Example: In a typical furane binder system:

B = 1.4% resin + 0.6% catalyst = 2.0%
T = 0.7 (only 30% of the binder residue is removed)
R = 0.90 (90% of reclaimed sand is reused with 10% new sand)

$$P = \frac{0.7 \times 2.0}{1 - (0.7 \times 0.9)}$$

= 3.78% (residual binder that builds up on the sand)

This represents an inefficient reclaimer. Ideally P should not exceed 3.0%. Even with an inefficient reclaimer P = 3% can be achieved by reducing R, that is, by adding more new sand. For example, reducing R to 0.75 (25% addition of new sand) reduces P to 2.95%.

Regular testing of reclaimed sand for LOI, acid demand, grain size and temperature is needed, together with regular maintenance of the reclaimer to ensure that consistent mould quality is achieved.

Binder systems containing inorganic chemicals, e.g. silicate-based systems, alkaline phenolic resins or binder systems containing phosphoric acid are difficult to reclaim at high percentages because no burnout of the inorganic material occurs.

Use of reclaimed sand with high LOI may cause problems due to excessive fumes at the casting stage, particularly if sulphonic acid-catalysed furane resins are used.

Thermal reclamation

Sand bonded with an entirely organic binder system can be 100% reclaimed by heating to about 800°C in an oxidising atmosphere to burn off the binder residues, then cooling and classifying the sand. Thermal reclaimers are usually gas heated but electric or oil heating can also be used. The steps in the process are:

Lump breaking
Metal removal
Heating to about 800°C for a certain time in a fluidised bed furnace or rotary kiln
Cooling the sand, using the extracted heat to preheat the incoming sand or the combustion air
Classification
Addition of new sand to make up losses in the system

Thermal reclamation is costly because of the large amount of heat needed and the relatively expensive equipment. The ever growing cost of sand disposal, however, has led to its increasing use.

Sand losses

Whatever method of reclamation is used, there is always some loss of sand so that 100% reclamation can never be achieved. Sand losses include: burn-on, spillage, inefficiencies in the sand system and the need to remove fines. Dust losses of around 5% can be expected and total sand losses of up to 10% may be expected.

Typical usage of sand reclamation

Furane bonded sand

Mechanical attrition allows up to 90% of sand to be reused. Only sulphonic acid-catalysed sand can be reclaimed. Reclaimed sand may have up to 3% LOI. Binder additions on rebonding can be reduced by 0.15–0.2% (from, say, 1.2% on new sand to 1.0%) with a proportionate reduction of catalyst. Nitrogen and sulphur levels build up and may cause casting defects and excessive fume if allowed to exceed 0.15% when the sand is rebonded. Low nitrogen resins must be used.

Thermal treatment at 800°C can be used on the residual sand. Nitrogen and sulphur are removed together with the remaining organic content.

Phenolic urethane bonded sand

Up to 90% of sand can be reclaimed by mechanical attrition. If iron oxide is added to the sand mix, care must be taken to ensure that the iron content of the sand does not rise too high or the refractoriness may suffer.

Thermal reclamation can be used on the residual sand as long as the iron oxide content is low.

Cold-box core sand, from core assembly casting processes, can be thermally reclaimed satisfactorily. If uncontaminated core lumps can be separated from green sand, they can also be reclaimed.

Alkaline phenolic bonded sand

These binders contain potassium which, if allowed to rise above 0.15% in the reclaimed sand, will cause unacceptable reduction of work time and final strength. This limits attrition reclamation to about 70%.

Thermal reclamation can be used on the residual sand if it is treated with FENOTEC ADTI, an anti-fusion additive. This additive prevents sintering of the sand during thermal treatment and aids removal of potassium salts.

Resin shell sand

Shell moulding and core sand (silica or zircon) is fully reclaimable by thermal means. The used sand often has as much as 2% of residual resin left after casting. Useful heat can be obtained from the combustion of the resin residues which significantly reduces the quantity of heat that must be supplied. The quality of the reclaimed sand must be regularly checked for:

> Size grading
> LOI (the visual appearance of the sand, which should be the colour of new sand, gives a good indication of LOI)
> Inorganic residues, e.g. calcium stearate additions, which are often made to lubricate shell sand, burn to CaO during reclamation, giving rise to high acid demand

Foundries wishing to thermally reclaim shell sand must have equipment to recoat the sand for reuse.

Silicate-ester bonded sand

Standard silicate-ester bonded sand can be reclaimed mechanically only at rather low levels, less than 50% because the build-up of soda in the sand reduces its refractoriness. The presence of reclaimed sand in the rebonded mix reduces the work time and final strength and increases the tendency for sagging.

VELOSET special silicate-ester binder

Up to 90% reclaimed sand can be reused with the VELOSET system. Shakeout sand is reduced to grain size in a vibratory crusher which provides the primary attrition stage. The sand is then dried in a fluidised bed drier. Secondary attrition takes place next in a hammer mill. The sand is finally passed through a cooler-classifier ready for reuse. The reclaimed sand is blended with new sand in the proportion 75 to 25. During the first 10 cycles of reuse, the sand system stabilises and the bench life of the sand increases by a factor of up to 2. Also, mould strength should improve, and it is usually possible to reduce the binder addition level by up to 20% yet still retaining the same strength as achieved using new sand. Once the process has become established, it becomes possible to reuse up to 85–90% of the sand, Figs 14.4 and 14.5.

Wet reclamation

Some silicate bonded sands are particularly difficult to reclaim mechanically because Na_2O builds up and lowers refractoriness. (VELOSET sand is an exception, see above.) Thermal reclamation is ineffective, but water washing can be used. The steps in the process are:

Lump breaking
Metal separation by magnet or screen
Disintegration to grain size
Water wash with mechanical agitation to wash off the silicate residues
Separation of sand and water
Drying of the sand
Agglomeration of the alkaline residues in the water to allow settlement and separation
Water treatment to permit safe disposal of the water

Wet reclamation is expensive and its use is not common.

Self-hardening resin binder systems

Furanes

Foseco products: FUROTEC, ESHANOL binders.

Principle: Self-setting furane sands use a furane resin and an acid catalyst. The resins are urea-formaldehyde (UF), phenol-formaldehyde (PF), or UF-PF resins with additions of furfuryl alcohol (FA). Speed of setting is controlled by the percentage of acid catalyst used and the strength of the acid.

Sand: Since the resins are acid catalysed, sands should have a low acid demand, less than 6 ml.

Resin: A wide range of resins is available having different nitrogen content. The UF base resin contains about 17–18% N; furfuryl alcohol is N free, so the N content of a UF-FA resin depends on its FA content. Nitrogen can cause defects in steel and high strength iron castings, so it is advisable to use high FA resins (80–95% FA, 3.5–1% N) although they are more expensive than lower FA resins. These resins are particularly useful with sands of high quality, such as the German Haltern and Frechen sands, which are round grained and can be used with low resin additions (<1.0%) so the total nitrogen on the sand is low.

A range of UF-PF furanes is also available, having low nitrogen content; they are frequently used with sands of rather lower quality, which require quite high resin additions (>1.2%). The total nitrogen on the sand can be kept to a lower level than is possible with UF-FA resins.

Phenol-formaldehyde resin is N free and a range of totally N-free PF-FA resins is available, these are used for particularly sensitive castings, or when sand is reclaimed. Where the presence of phenol in the resin presents a problem, due to restrictions on used sand disposal, special phenol-free FA, Resorcinol (FA-R) resins can be used. They contain less than 0.5% N and are used for steel and high quality iron casting. While they are expensive, they are used at low addition rates.

Catalyst: UF-FA resins can be catalysed with phosphoric acid, which has the advantage of low odour and no sulphur but reclamation of the sand is difficult. FA-R, PF-FA, UF-FA with high FA and UF-PF-FA resins are catalysed with sulphonic acids, either PTSA (para-toluene sulphonic acid), XSA (xylene sulphonic acid) or BSA (benzene sulphonic acid), the last two are strong acids and are used when faster setting is required. Sulphonic acid-catalysed resin sand can be reclaimed but the sulphur content of the sand rises and may cause S defects in ferrous castings particularly in ductile iron. Environmental problems may also arise due to the SO_2 gas formed when moulds are cast. Mixed organic and inorganic acids may also be used.

Addition rate: Resin: 0.8–1.5% depending on sand quality. Catalyst: 40–60% of resin, depending on sand temperature and speed of setting required. If fast setting is required, XSA or BSA catalyst should be used.

Pattern equipment: Wood or resin is commonly used. Metal patterns can be used but if the pattern is cold, setting may be retarded. FA is a powerful solvent and will attack most paints; patterns must be clean and preferably unpainted. Special release agents such as STRIPCOTE AL can be used.

Temperature: The optimum is 20–30°C. Cold sand seriously affects the setting speed and the final strength.

Strength: Depends on resin type and addition, typically 4000 kPa (600 psi) compression strength.

Speed of strip: Can be from 5 to 30 minutes. Short strip times require high speed mixers and may cause problems of sand build-up on the mixer blades.

Work time/strip time: UF-FA resins are better than PF-FA; the higher the FA content, the better.

Coatings: Water or spirit-based coatings may be used; alcohol-based coatings may cause softening of the mould or core surface and it is advisable to allow $\frac{1}{2}$–1 hour between stripping and coating.

Environmental: Formaldehyde is released from the mixed sand, so good extraction is needed around the mixer. When sulphonic acid catalyst is used, SO_2 gas is evolved when moulds are cast, particularly if reclaimed sand is used.

Reclamation: Attrition reclamation works well and it is often possible to reduce the resin addition on reclaimed sand. Low N resin and sulphonic acid catalyst must be used. The LOI of the reclaimed sand must be kept to less than 3.0%. N and S in the sand should not exceed 0.15% when the sand is rebonded.

Casting characteristics: N above 0.15% in the sand will cause defects in iron and steel castings. S above 0.15% in the sand may cause reversion to flake graphite at the surface of ductile iron castings. Metal penetration can be a problem if work time is exceeded.

General: Probably the most widely used self-hardening system because of the easy control of set times, good hot strength, erosion resistance and the ease of reclamation.

Phenolic-isocyanates (phenolic-urethanes)

Foseco product: POLISET binder.

Principle: The binder is supplied in three parts. Part 1 is a phenolic resin in an organic solvent. Part 2 is MDI (methylene diphenyl diisocyanate). Part 3 is a liquid amine catalyst. When mixed with sand, the amine causes a reaction between resin and MDI to occur, forming urethane bonds which rapidly set the mixture. The speed of setting is controlled by the type of catalyst supplied in Part 1.

Sand: The binder is expensive, so good quality sand is needed to keep the cost of additions down. AFS 50–60 is usually used.

Addition rate: The total addition is typically 0.8% Part 1, 0.5% Part 2, more or less being used, depending on the sand quality.

Pattern equipment: Wood, resin or metal can be used. Paints must be resistant to the strong solvents in Part 1 and Part 2.

Temperature: Sand temperature affects cure rate, but not as seriously as other binders. The optimum temperature is 25–30°C.

Speed of strip: Can be very fast, from 2 to 15 minutes. A high speed mixer is needed to handle the fast cure rates.

Work time/strip time: Very good.

Strength: Compression strength is typically over 4000 kPa (40 kgf/cm^2, 600 psi).

Coatings: If water-based coatings are used, they should be applied immediately after strip and dried at once. Spirit-based coatings should be applied 15–20 minutes after strip and are preferably air dried.

Casting characteristics: Lustrous carbon defects may occur on ferrous castings. Part 2 (MDI) contains 11.2% N and gas defects may occur in steel castings. The binder has rather low initial hot strength, so erosion defects may occur. All these problems can be reduced by the addition of 1 or 2% of iron oxide to the sand.

Reclamation: Attrition works well, 90% reuse of sand is possible. If iron oxide is added to the sand mix, care must be taken to ensure that the iron content of the sand does not rise too high or the refractoriness may suffer.

Environmental: Solvent release during mixing and compaction may be troublesome, use good exhaust ventilation. Isocyanates can cause respiratory problems in sensitised individuals. Once an individual has become sensitised, exposure, however small, may trigger a reaction. MDI has very low vapour pressure at ambient temperature, so exposure to the vapour is unlikely to cause problems, however, uncured isocyanate may be present on airborne sand particles at the mixing station. Good exhaust ventilation is essential.

General: This system has never been as popular in the UK and Europe as in the USA due to the difficulty of stripping and the low hot strength. Steel foundries find it useful for cores, where the low hot strength reduces hot tearing.

Alkaline phenolic resin, ester hardened

Foseco products: FENOTEC, FENOTEC hardener.

Principle: The binder is a low viscosity, highly alkaline phenolic resole resin. The hardener is a liquid organic ester. Sand is mixed with hardener and resin, usually in a continuous mixer. The speed of setting is controlled by the type of ester used.

Sand: Can be used with a wide range of sands including zircon, chromite and high acid demand sand such as olivine.

Resin addition: 1.2–1.7% depending on the sand quality; 18–25% hardener based on resin.

Nitrogen content: Very small or zero.

Pattern equipment: Wood or resin, patterns strip well.

Temperature: Low sand temperature slows the cure rate, but special hardeners are available for cold and warm sand.

Speed of strip: 3 minutes to 2 hours depending on the grade of hardener used. The work time/strip time ratio is good.

Strength: 24 hr strength:	transverse	1600 kPa	16 kgf/cm^2	230 psi
	tensile	900	9	130
	compression	4000	40	600

Coatings: Some iron and steel castings can be made without coatings. Both water- and spirit-based coatings can be used, but some softening-back of the sand surface is possible. The stripped core or mould should be allowed to harden fully before applying the coating. Water-based coatings should be dried as quickly as possible. Alcohol-based coatings should be fired as soon as possible after application.

Casting characteristics: Good as-cast finish on all metals. Hot tearing and finning defects are eliminated. No N, S or P defects. Good breakdown, particularly on low melt point alloys. Widely used for steel castings as well as iron and aluminium.

Reclamation: FENOTEC binders allow up to 70–90% sand reclamation; there is some loss of strength and careful management of the alkali content of the reclaimed sand is needed. Thermal reclamation can be used if the sand is treated with FENOTEC ADTI, an anti-fusion additive which aids potassium removal.

Environmental: Low fume evolution at mixing, casting and knockout stages.

General: Stripping from all pattern types is excellent; this, together with the good casting finish achieved and the low fume evolution, make the system popular for all types of casting but most of all for steel.

Triggered hardening systems

Cores for repetition foundries are usually made using a triggered hardening system with the mixed sand being blown into the core box. The cores must be cured in the box until sufficient strength has been achieved to allow stripping without damage or distortion; usually the core continues to harden after stripping. Transverse breaking strength or tensile strength are used to assess the properties of triggered core binder systems since they represent the properties needed to strip and handle cores. The strength requirements needed depend on the particular type of core being made. Thin section cores, such as cylinder head water jacket cores, require high stripping strength because of their fragile nature. Tensile strengths of 1000–2000 kPa (150–300 psi) are typical, equating roughly to transverse strengths of 1500–3000 kPa. Final strengths may be higher, but some binder systems are affected by storage conditions (humidity in particular) and strengths may even fall if the storage conditions are poor. Surface hardness is also important but difficult to measure precisely.

Heavy section cores and moulds made by core blowing can be successfully made with lower strength binders, and factors such as fume released on casting and compatibility of core residues with the green and system may be more important than achieving the very highest strength.

Heat triggered processes

The sand and binder are mixed then introduced into a heated core box or pattern. The heat activates the catalyst present in the binder system and cures the binder. Core boxes and patterns must be made of metal, normally grey cast iron, evenly heated by means of gas burners or electric heating elements. The working surface of the core box is usually heated to 250°C, higher temperatures may overcure or burn the binder. Sand is a rather poor conductor of heat and heat penetration is rather slow, Fig. 13.4; it is therefore difficult to cure thick sections of sand quickly and fully. When the core is ejected from the core box, residual heat in the sand continues to penetrate into the core, promoting deeper cure. To achieve the fastest cure times, heavy section cores are often made hollow, using heated mandrels or "pull-backs" to reduce core thickness. Core boxes are usually treated with silicone release agents to improve the strip. The thermal expansion of the metal core box must be taken into account when designing core boxes. A cast iron core box cavity 100 mm long will expand by 0.27 mm when heated from 25 to 250°C. This change becomes significant on large cores.

Other methods of applying heat to sand cores have been tried. Microwave or dielectric heating is difficult because electrically conducting metal core boxes cannot be used. Certain resins can be used for core boxes but they pick up heat from the cores and may distort.

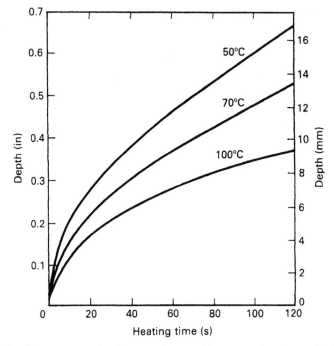

Figure 13.4   *Temperature rise in a sand core with one surface in contact with a heated core box at 250°C (theoretical).*

Blowing heated air through the core has also been used, but a large volume of air is needed and the method becomes slow and impractical if core sections above about 30 mm are to be cured. Approximately 1 kg (800 litres at STP) of heated air is needed to heat 1 kg of sand to curing temperature.

Gas triggered systems

Sand and binder are mixed and blown into a core box, then a reactive gas is blown in causing hardening of the binder. Hardening occurs at room temperature, avoiding the need for heated core boxes. Many of the gases used to trigger chemical reactions are highly reactive and toxic so that specially designed gas generators are needed to meter the gas accurately into the core box. Core boxes must be sealed to contain the gases and often scrubbers must be used to absorb the exhaust gas to prevent it from contaminating the atmosphere both inside and outside the foundry. Frequently, the gassing cycle includes an air purge to remove residual gas from the core so that it can be handled and stored safely.

Gas cured binder systems often allow very fast curing, even of thick section cores.

Heat triggered processes

The shell or Croning process

Principle: Sand is precoated with a solid phenolic novolak resin and a catalyst to form a dry, free-flowing material. The coated sand is blown into a heated core box or dumped onto a heated pattern plate causing the resin to melt and then harden. Cores may be solid or they may be hollow with the uncured sand shaken out of the centre of the core. Shell moulds are normally only 20–25 mm thick.

Sand: The base sand is usually a fine silica sand of AFS 60–95. To achieve the highest strength with the minimum resin addition, the sand should be pure, rounded grain and free from surface impurities. Zircon sand is also precoated for special applications.

Resin and catalyst additions: The resins are solid phenolic novolaks which melt at around 100°C. The catalyst is hexamine, a white powder. Resin additions are 2.5–4.5% (depending on the application and the strength required), hexamine is added at 11–14% of the resin content.

Nitrogen content: The resins are N free but hexamine contains 40% N. Special low N systems with little or no hexamine are available.

Precoating procedure: There are two processes: hot coating and warm coating. In the hot coating procedure, the sand is heated to around 130°C and charged into a batch mixer. Solid resin granules are added, the heat from the sand melts the resin which then coats the sand grains. Aqueous hexamine solution plus release agents are added, cooling the sand below the melt point of the resin. The coated sand is broken down to grain size and finally cooled.

In warm coating, the sand is heated to 50°C and charged into a batch mixer together with hexamine and release agents. Resin in alcohol solution is added and warm air blown through the mixer to vaporise the solvent, leaving the sand evenly coated. After breaking down the sand lumps and cooling, the coated sand is ready for use.

Many foundries purchase ready coated sand from specialist suppliers.

Bench life: Precoated sands last indefinitely if stored dry and not exposed to excessive heat, which may cause clumping.

Core blowing: The free flowing sand can be blown at low pressure, 250–350 kPa (35–50 psi). Core boxes should be made of cast iron heated to 250°C. Copper or brass must be avoided since ammonia released during curing will cause corrosion.

Curing times: Minimum curing time is 90 seconds but 2 minutes is common. The longer the curing time, the thicker is the shell build-up. Increasing curing temperature does little to speed up cure rate and runs the risk of over-curing the sand adjacent to the core box or pattern plate.

Core strength: Surface hardness and strength is high on ejection. A 3.5% resin content will give a tensile strength of 1400 kPa (200 psi). Storage properties of cured cores is excellent.

Gas evolution: 4–5 ml/g of gases for each 1% of resin.

Casting characteristics: Core and mould coatings are unnecessary. The surface finish of castings is excellent. Breakdown is good, particularly with hollow cores. Accuracy and reproduction of pattern detail is extremely good because of the free-flowing properties of precoated sand. Some distortion, particularly of moulds, can occur on casting and support of shell moulds by steel shot or loose sand may be needed for heavy section castings. N defects may occur on steel castings unless low N coated sand is used. Excellent surface finish is achieved on aluminium castings.

Health hazards: Phenol and ammonia are released during curing so good ventilation is needed at core- and mould-making machines. Casting into shell moulds produces unpleasant and hazardous fumes and often shell moulds are cooled in an enclosed tunnel, allowing fumes to be collected and exhausted outside the foundry.

Reclamation: When both moulds and cores are made by shell process, the sand can be reclaimed thermally.

Shell moulding: Shell moulds are made by dumping precoated sand onto an iron pattern plate heated to 240–260°C. After a suitable time, usually about 2 minutes, the mould is overturned, returning the uncured sand to the hopper and leaving a shell mould 20–25 mm thick which is ejected from the pattern plate. Cores are placed and the two half-moulds are glued together with hot-melt adhesive (CORFIX).

Shell moulds are usually cast horizontally, unsupported. If cast vertically, support is usually necessary to prevent mould distortion. Steel shot is frequently used to support the mould.

KALMIN pouring cups are often used with horizontally poured shell moulds.

General: The shell process was one of the first synthetic resin foundry processes to be developed. Although it is slow and rather expensive, it is still widely used, because of the excellent surface finish and dimensional accuracy of the castings produced.

Hot-box process

Principle: The binder is an aqueous PF-UF or UF-FA resin, the catalyst is an aqueous solution of ammonium salts, usually chloride and bromide. Sand is mixed with the liquid resin and catalyst and blown into a heated core box. The heat liberates acid vapour from the catalyst which triggers the hardening reaction. Hardening continues after removal of the core from the box causing the sand to be cured throughout.

Sand: Clean silica sand of AFS 50–80, acid demand should be low. The binder is aqueous so traces of moisture in the sand do not affect the cure. Low sand temperatures (below 20°C) slow the cure rate.

Resin addition: 2.0–2.5% depending on sand quality. Catalyst additions are 20–25% of the resin weight.

Nitrogen content: The resin contains 6–12% N. The catalyst are usually ammonia chloride solutions and also contain N. The total N in the system is 10–14% depending on the resin used. Low N or N-free systems are available but their performance is not as good as the N-containing binders.

Mixing procedure: Batch or continuous mixers can be used; add catalyst first then the resin.

Bench life: 1–2 hours, mixed sand must be kept in closed containers to prevent drying out. High ambient temperatures reduce bench life.

Core blowing: High blowing pressure, 650–700 kPa (90–100 psi) is needed to blow the wet sand into core boxes.

Core boxes: Cast iron or steel, fitted with gas or electric heaters to maintain the temperature at 250–300°C. Silicone parting agents are usually applied to improve the strip of the core.

Curing time: Thin section cores cure in 5–10 seconds. As cores increase in section, curing time must be extended up to about 1 minute for 50 mm section. For larger sections still, it is advisable to lighten out cores with heated mandrels, otherwise the centre of the core will remain uncured.

Core strength: Surface hardness and strength are high on ejection. Final tensile strength is 1400–2800 kPa (200–400 psi). Storage properties are good if cores are kept dry.

Casting characteristics:

Ferrous castings:	Cores are usually coated to prevent burn-on. Water-based coatings, such as RHEOTEC, are used. The

hot strength and breakdown is good, particularly for the UF-PF-FA resins, the higher the PF content, the worse the breakdown. The rather high nitrogen can cause gas holes and fissures in iron castings and additions of 1–3% of a coarse-grained form of iron oxide are often used to minimise N defects. Finely powdered red iron oxide can also be used but it causes some loss of strength.

Hot-box cores made using phenolic resin expand slightly during casting and this can cause distortion, for example on large water jacket cores. Iron oxide additions reduce the expansion, so does the use of UF-FA resins.

Aluminium castings: Special hot-box resins are available which break down readily at aluminium casting temperatures.

Copper-based castings: Brass water fittings are commonly made with special hot box resin cores.

Health hazards: Formaldehyde is liberated during curing, so core blowers and the area around them must be well ventilated and exhausted. Avoid skin contact with liquid resins and mixed sand.

Reclamation: It is not normal to reclaim hot-box cores when they are used in green sand moulds. Core sand residues in green sand are well tolerated.

General: The hot-box process was the first high speed resin core-making process. Gas-cured cold-box processes have superseded it in many cases, particularly for thick section cores, but hot box is still used because of its good surface hardness, strength in thin sections and excellent breakdown properties.

Warm-box process

Principle: The binder is a reactive, high furfuryl alcohol binder. The catalyst is usually a copper salt of sulphonic acid. Sand, binder and catalyst are mixed and blown into a heated core box. The heat activates the catalyst which causes the binder to cure.

Sand: Clean silica sand of AFS 50–60 is used. Low acid demand is advisable.

Resin addition: 1.3–1.5% resin and 20% catalyst (based on resin) depending on sand quality.

Nitrogen content: The resin contains about 2.5% N. The catalyst is N free.

Mixing procedure: Continuous or batch mixers. Catalyst should be added first, then resin.

Bench life: Typically 8 hours.

Core blowing: The binder has low viscosity and blow pressures of around 500 kPa (80 psi) may be used.

Core boxes: Cast iron or steel heated to 180–200°C. Use a silicone-based parting agent. A release agent is sometimes added to the sand/binder mix.

Curing time: 10–30 seconds depending on the thickness. Thick section cores continue to cure after ejection from the core box.

Core strength: Surface hardness and strength is high on ejection. Final tensile strength can be 3000–4000 kPa (400–600 psi).

Casting characteristics: Gas evolution is low, the low nitrogen content reduces incidence of gas-related defects in ferrous castings. Surface finish of castings is good with low incidence of veining defects. Breakdown after casting is good.

Environmental: Emission of formaldehyde is low at the mixing and curing stages. Avoid skin contact with binder, catalyst and mixed sand.

Reclamation: Core residues in green sand are not harmful.

General: Warm-box cores have high strength and resistance to veining and are used in critical applications, usually for iron castings such as ventilated brake discs.

Oil sand

Principle: Certain natural oils, such as linseed oil, known as "drying oils", polymerise and harden when exposed to air and heat. Natural oils can be chemically modified to accelerate their hardening properties. Silica sand is mixed with the drying oil, a cereal binder and water. The resulting mixture is either manually packed or blown into a cold core box. The cereal binder, or sometimes dextrin, gives some green strength to the core which is then placed in a shaped tray or "drier" to support it during baking. The backing hardens the oil and the core becomes rigid and handleable.

Sand: Clean silica sand, AFS 50–60.

Binder addition: 1–2% of drying oil
1–2% pre-gelatinised starch ⎫ or 2% dextrin
2–2.5% water ⎭ 0.5–1% water

Nitrogen content: Zero.

Mixing: Batch mixers are preferred in order to develop the green bond strength, mixing times may be 3–10 minutes.

Bench life: As long as the mixture does not lose moisture, the bench life is up to 12 hours.

Core blowing: Oil sand mixtures are sticky and difficult to blow, the highest blow pressure possible (around 700 kPa, 100 psi) is used. They are frequently hand-rammed into core boxes.

Core boxes: May be wood, plastic or metal. Wooden boxes require a paint or varnish to improve the strip. Metal boxes are preferably made of brass or bronze to aid stripping of the fragile green cores.

Curing: A recirculating air oven is needed since oxygen is necessary to harden the oil. The temperature is normally 230°C, allowing 1 hour for each 25 mm section thickness. Burning and consequent friable edges may occur on thin section cores; if this happens, a lower temperature for a longer time should be used.

Core strength: Green strength is low so sagging will occur if the cores are not supported during baking. Correctly baked cores develop tensile strength of 1340 kPa (200 psi).

Casting characteristics: Breakdown after casting is excellent. The gas evolution is high, particularly if cores are under-baked, venting is often necessary. Water-based coatings can be used. While for most applications, oil sand has been superseded by synthetic resin processes, it still remains a valuable process in applications where particularly good breakdown of cores is needed, for example for high conductivity copper and high silicon iron castings in which hot tearing is a serious problem.

Health hazards: Unpleasant fumes are emitted during baking and ovens must be ventilated. Some proprietary core oils contain a proportion of mineral oil, which may be harmful if skin contact is prolonged.

Gas triggered processes

Phenolic-urethane-amine gassed (cold-box) process

Foseco products: POLITEC and ESHAMINE cold-box resin.

Principle: The binder is supplied in two parts. Part 1 is a solvent-based phenolic resin, Part 2 is a polyisocyanate, MDI (methylene di-phenyl di-isocyanate) in a solvent. The resins are mixed with sand and the mixture blown into a core box. An amine gas (TEA, triethylamine, or DMEA, dimethyl ethyl amine) is blown into the core, catalysing the reaction between Part 1 and Part 2 causing almost instant hardening.

Sand: Clean silica sand of AFS 50–60 is usually used but zircon and chromite sands can be used. The sand must be dry, more than 0.1% moisture reduces the bench life of the mixed sand. High pH (high acid demand) also reduces bench life. Sand temperature is ideally about 25°C; low temperature causes amine condensation and irregular cure. High temperature causes solvent loss from the binder and loss of strength.

Resin addition: Total addition is 0.8–1.5% depending on sand quality. Normally equal proportions of Part 1 and Part 2 are used.

Nitrogen content: Part 2, the isocyanate, contains 11.2% N.

Mixing procedure: Batch or continuous mixers can be used, add Part 1 first then Part 2. Do not over-mix since the sand may heat and lose solvents.

Bench life: 1–2 hours if the sand is dry.

Core blowing: Use low pressure, 200–300 kPa (30–50 psi), the blowing air must be dry, use a dessicant drier to reduce water to 50 ppm. Sticking of sand and resin to the box walls can be a problem due to resin blow-off. The lowest possible blow pressure should be used. Use of a special release agent such as STRIPCOTE is advised.

Core boxes: Iron, aluminium, urethane or epoxy resin can be used. Wood is possible for short runs. Use the minimum vents that will allow good filling, since reduced venting gives better catalyst gas distribution. The exhaust vent area should be 70% of the input area to ensure saturation of the core. Core boxes must be sealed to allow the amine catalyst gas to be collected.

Gas generators: The amine catalysts are volatile, highly flammable liquids. Special generators are needed to vaporise the amine and entrain it in air or CO_2. The carrier gas should be heated to 30–40°C to ensure vaporisation. Controlled delivery of amine by pump or timer is desirable.

Gas usage: Approximately 1 ml of amine (liquid) is needed per kilogram of sand. The amine usage should be less than 10% of Part 1 resin. DMEA is faster curing than TEA. Typical curing is a short gassing time (1–2 seconds) followed by a longer (10–20 seconds) air purge to clear residual amine from the core. Over-gassing is not possible but simply wastes amine.

Typical gassing times:

core wt	total gas + purge
10 kg	10 sec.
50	30
150	40

Core strength: Tensile strength immediately after curing is high, 2000 kPa (300 psi), transverse strength 2700 kPa (400 psi). Storage of cores at high humidity reduces the strength considerably. The use of water-based coatings can cause loss of strength.

Casting characteristics:

| Ferrous castings: | Good surface and strip without coatings. Low hot strength but this can be improved by adding iron oxide. Tendency to finning or veining. Lustrous carbon formation may cause laps and elephant skin defects on upper surfaces of castings. Addition of red iron oxide (0.25–2.0%) or a coarse-grained from of black iron oxide at 1.0–4.0% reduces defects. Breakdown is good. The N content may cause problems on some steel castings |

Ferrous castings: Good surface and strip without coatings.
Low hot strength but this can be improved by adding iron oxide
Tendency to finning or veining
Lustrous carbon formation may cause laps and elephant skin defects on upper surfaces of castings. Addition of red iron oxide (0.25–2.0%) or a coarse-grained from of black iron oxide at 1.0–4.0% reduces defects
Breakdown is good
The N content may cause problems on some steel castings

Aluminium castings: Good surface and strip
Poor breakdown, the resin hardens when heated at low temperature. Aluminum castings may need heat treatment at 500°C to remove cores
There are no hydrogen problems

Reclamation: Excessive contamination of green sand with cold-box core residues can cause problems. Cold-box cores and moulds can be thermally reclaimed if uncontaminated with iron oxide.

Environmental: TEA and particularly DMEA have objectionable smells even at 3 or 6 ppm. Good, well-sealed core boxes, good exhaust and good exhaust gas scrubbers are necessary. The cores must be well purged or amine will be released on storage. Liquid amine is highly flammable, treat like petrol. Air/amine mixtures may be explosive. MDI (Part 2) acts as a respiratory irritant and may cause asthmatic symptoms but it has low volatility at ambient temperature and is not normally a problem. Avoid skin contact with resins or mixed sand.

General: In spite of its environmental and other problems, the cold-box process is so fast and produces such strong cores that it is the most widely used gas triggered process for high volume core production. Good engineering has enabled the environmental problems to be overcome. Reduced free-phenol resins are being produced to assist with sand disposal problems.

ECOLOTEC process (alkaline phenolic resin gassed with CO_2)

Foseco product: ECOLOTEC resin.

Principle: ECOLOTEC resin is an alkaline phenol-formaldehyde resin containing a coupling agent. The resin is mixed with sand and the mixture blown into a core box. CO_2 gas is passed through the mixture, lowering the pH and activating the coupling agent which causes crosslinking and hardening of the resin. Strength continues to develop after the core is ejected as further crosslinking occurs and moisture dries out.

Sand: Clean silica sand of AFS 50–60 is used. The sand should be neutral (pH7) with low acid demand. Zircon and chromite sands can be used, but olivine sand is not suitable. Temperature is ideally 15–30°C.

Resin addition: 2.0–2.5% depending on the sand grade.

Nitrogen content: Zero.

Mixing procedure: Batch or continuous.

Bench life: Curing occurs only by reaction with CO_2. The CO_2 in the air will cause a hardened crust on the surface of the mixed sand, but the soft sand underneath can be used. Keep the mixed sand covered. The bench life is usually 5 hours at 15°C reducing to 1 hour at 30°C.

Core blowing: Blow pressure 500–700 kPa (80–100 psi) is needed.

Core boxes: Wood, metal or plastic. Clean them once per shift and use SEPAROL or a silicone-based release agent.

Gassing: Cores are blown at 400–550 kPa (60–80 psi) then gassed for 20–40 seconds at 100–300 litres/minute. CO_2 consumption is about 2% (based on weight of sand). Gas should not be forced through the core box at high velocity since the gas must react with the binder. No purge is necessary before extracting the core.

Core strength:

	as gassed		
	kPa	*kgf/cm²*	*psi*
Compression	2000–3000	20–30	300–400
Tensile	500–800	5–8	70–110
Transverse	1000–1500	10–15	145–210

Cores should be stored in dry conditions.

Casting characteristics: Steel, iron, copper-based and light alloy castings can be made. ECOLOTEC is free from N, S and P. Finning and lustrous carbon defects do not occur. Breakdown after casting is good.

Coating: Water or solvent-based coatings can be used without affecting the strength. Methanol coatings should be avoided.

Environmental: Gassed cores have no odour. Fume is low at mixing, casting and knockout.

Reclamation: Not normally practised when cores are used in green sand moulds. Core sand residues entering the green sand cause no problems.

General: Tensile strength is not as high as the amine gassed isocyanate process but the excellent casting properties, freedom from nitrogen, lustrous carbon and finning defects and above all its environmental friendliness make ECOLOTEC an attractive process particularly for larger cores and moulds for iron, steel and aluminium castings.

The SO_2 process

Principle: Sand is mixed with a furane polymer resin and an organic peroxide, the mixture is blown into the core box and hardened by passing sulphur dioxide gas through the compacted sand. The SO_2 reacts with the peroxide forming SO_3 and then H_2SO_4 which hardens the resin binder.

Sand: Clean silica sand of AFS 50–60. Other sands may be used if the acid demand value is low. The temperature should be around 25°C, low temperature slows the hardening reaction.

Resin addition: Typically 1.2–1.4% resin, 25–60% (based on resin) of MEKP (methyl ethyl ketone peroxide).

Nitrogen content: Zero.

Mixing: Batch or continuous mixers, add resin first then peroxide. Do not overmix since the sand may heat and reduce bench life.

Bench life: Up to 24 hours.

Core blowing: Blowing pressure 500–700 kPa (80–100 psi).

Core boxes: Cast iron, aluminium, plastics or wood can be used. A build-up of resin film occurs on the core box after prolonged use, so regular cleaning is needed by blasting with glass beads or cleaning with dilute acetic acid.

Curing: The SO_2 gas is generated from a cylinder of liquid SO_2 fitted with a heated vaporiser. The gas generator must also be fitted with an air purge system so that the core can be cleared of SO_2 before ejection. SO_2 is highly corrosive and pipework must be stainless steel, PTFE or nylon. For large cores, a separate gassing chamber may be used in which the chamber pressure is first reduced then SO_2 is injected and finally the chamber is purged.

Gas usage: 2 g (ml) liquid SO_2 is needed per kg of sand. Cores are normally gassed for 1–2 seconds followed by 10–15 seconds air purge. Overgassing is not possible.

Core strength: Tensile strength is 1250 kPa (180 psi) after 6 hours. Storage properties are good.

Casting characteristics: Coatings are not normally necessary and surface quality of castings is good. Breakdown of cores is excellent with ferrous castings. It is also good with aluminium castings, better than cold-box cores, and this has proved to be one of the best applications. The sulphur catalyst may cause some metallurgical problems on the surface of ductile iron castings.

Reclamation: Thermal reclamation is possible. Mechanically reclaimed sand should not be used for SO_2 core making but may be used with furane self-setting sand.

Environmental: SO_2 has an objectionable smell and is an irritant gas. Core boxes must be sealed and exhaust gases must be collected and scrubbed with sodium hydroxide solution. Cores must be well purged to avoid gas release during storage. MEKP is a strongly oxidising liquid and may ignite on contact with organic materials. Observe manufacturer's recommendations carefully. Avoid skin contact with resin and mixed sand.

SO₂-cured epoxy resin

Principle: Modified epoxy/acrylic resins are mixed with an organic peroxide, the mixture is blown into the core box and hardened by passing sulphur dioxide gas through the compacted sand. The SO_2 reacts with the peroxide forming SO_3 and then H_2SO_4 which hardens the resin binder.

Sand: Clean silica sand of AFS 50–60. Other sands may be used if the acid demand value is low. The temperature should be around 25°C low temperature slows the hardening reaction.

Resin addition: Typically 1.2–1.4% resin, 25–60% (based on resin) of MEKP (methyl ethyl ketone peroxide).

Nitrogen content: Zero.

Other details: Similar to SO_2/furane process. Bench life is up to 24 hours.

Ester-cured alkaline phenolic system

Foseco product: FENOTEC resin.

Principle: The resin is an alkaline phenolic resin (essentially the same as the self-hardening resins of this type). Sand is mixed with the resin and blown or manually packed into a core box. A vaporised ester, methyl formate, is passed through the sand, hardening the binder.

Sand: Highest strengths are achieved with a clean, high silica sand of AFS 50–60. Sand temperature should be between 15 and 30°C.

Resin/peroxide addition: Typically 1.5% total.

Nitrogen content: Zero, sulphur is also zero.

Mixing procedure: Batch or continuous.

Bench life: Curing only occurs by reaction with the ester hardener, so the bench life is long, 2–4 hours.

Core blowing: Blow pressure 350–500 kPa (50–80 psi) is needed.

Core boxes: Wood, plastic or metal. Clean once per shift and use SEPAROL or a silicone-based release agent.

Gassing: Methyl formate is a colourless, highly flammable liquid, boiling at 32°C. It has low odour. A specially designed vaporiser must be used to generate the methyl formate vapour. Usage of methyl formate is about 20–30% of the resin weight. Core boxes and gassing heads should be sealed correctly and the venting of the core box designed to give a slight back pressure so that the curing vapour is held for long enough for the reaction to take place.

Core strength: Compression strengths of 5000 kPa (700 psi) are possible. Tensile and transverse strengths are not available, they are not as high as phenolic isocyanate resins.

Casting characteristics: Finning and lustrous carbon defects are absent. Good high temperature erosion resistance and good breakdown. Mostly used for steel and iron castings.

Environmental: Low odour, but methyl formate is flammable and care is needed.

Reclamation: Reclamation by attrition is possible but the reclaimed sand is best reused as a self-hardening alkaline-phenolic sand where strengths are not as critical. Core sand residues entering green sand causes no problems.

General: Strengths are relatively low so the process has mainly been used for rather thick section moulds and cores where high handling strength is not necessary.

Review of resin core-making processes

The heat-activated resin core-making processes, Croning shell and hot box were developed in the 1950s and 1960s and rapidly supplanted the older oil sand coremaking process. The attraction was speed of cure and the fact that cores could be cured in the box, eliminating the dimensional problems of oil sand cores. By the mid-1970s the amine-urethane cold-box process was becoming firmly established, bringing great benefits of speed as well as an environmental burden. Since 1976 there have been many further developments in the amine-urethane systems and in other cold gas-hardened core binder systems.

Reasons for the intense interest in gas hardened processes include:

The flexibility of gas hardened processes which makes them suitable for mass production of cores with automatic equipment
Fast curing of cores in the box by controlled injection of reactive gases
High strengths on ejection, often 80% of the final strength
Dimensional accuracy resulting from cold curing

Table 13.1   Core-making processes used in 48
automobile foundries in Germany, 1991 (Foseco
data)

Amine cold box	44
Hot box	10
Shell/Croning	9
CO_2-silicate	3

Availability of an extensive range of specially designed core making
equipment
Energy saving through cold operation
Rapid tooling changes made easy by low tooling temperatures

By 1990 the majority of cores were produced by the cold-box, amine
isocyanate process, Table 13.1.

Tables 13.2a and b summarise the production features of the many core-
making processes in use.
   Figure 13.5 shows how the use of chemical binders has changed in the
USA over the last 40 years:

Shell sand is declining slightly, holding up remarkably for such an old
process
Core oil has declined in the face of hot box and cold box

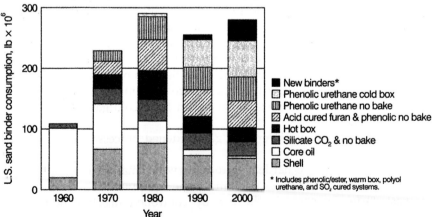

Figure 13.5   *US foundry industry consumption of sand binder by system. (From*
Foundry Management & Technology, *Jan. 1995, p. D-5.)*

Table 13.2a Production features of core binders

Process	Core-box requirements	Curing	Typical binder (%)	Flowability	Breakdown	Integration into green sand
Shell	Metal, heated	260°C	2–5	1	2	2
Hot box	Metal, heated	230°C	1.2–2.5	2	1	2
Warm-box	Metal, heated	150–180°C	1.2–2.0	2	1	2
Amine-urethane	Any material, vented, sealed	Amine Gas	0.9–2.0	1	1	2
SO_2-epoxy	Any material, vented, sealed	SO_2 gas and peroxide	1.0–1.2	1	1	2
SO_2-furane	Any material, vented, sealed	SO_2 gas	1.0–1.5	1	1	2
Methyl-formate Alkaline phenolic	Any material, vented, sealed	Methyl formate gas	1.4–1.6	1	2	1
CO_2-silicate	Any, vented	CO_2 gas	2.5–3.5	3	3	1
ECOLOTEC	Any, vented	CO_2 gas	2.0–2.5	1	2	1

1 Good – unlikely to restrict application
2 Satisfactory – suitable for most applications
3 Restrictions – may limit applications

Table 13.2b Properties of core binders

Process	Strength	Gas evolution properties	Typical bench life	Typical strip time	Minimum strip to pour time	Pin-hole tendency	Veining	Lustrous carbon
Shell	1	3	Indefinite	2 mins.	30 min	3	1	1
Hot box	1	2	4–6 hrs	1/2–1 min.	1/2–2 hrs	3	2	1
Warm box	1	1	4–8 hrs	1/2–1 min.	0–2 hrs	3	1	1
Amine-urethane	1	1	1–6 hrs	10–30 sec.	0–1 hr	2	2	2
SO_2-epoxy	1	2	24 hrs	10–30 sec.	Instant	1	1	1
SO_2-furane	2	2	24 hrs	10–30 sec.	Instant	2	2	2
CO_2-silicate	3	1	2–3 hrs	10–60 sec.	1 hr	1	1	1
ECOLOTEC	2	1	3–4 hrs	5–30 sec.	Instant	1	1	1

1 Good – unlikely to restrict application
2 Satisfactory – suitable for most applications
3 Restrictions – may limit applications

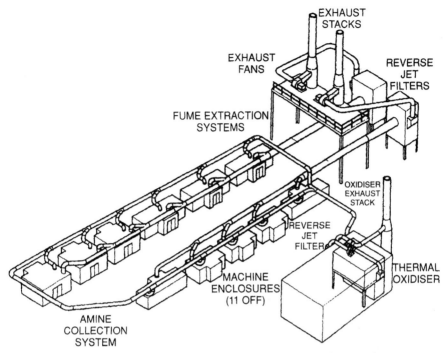

Figure 13.6 *Collection and control of amine gas in a cold-box core shop.*

Silicates received a boost in 1980, due to the introduction of ester
hardening, but have declined slowly since
Hot box has declined due to the rise in use of cold-box phenolic
urethane core binders, but still has a significant place
Acid cured furanes have held steady since the maximum in 1980
Urethane no-bakes are popular in the US, although their use in Europe
is limited
New binders are predicted to increase substantially by the year 2000

The environmental difficulties of the amine-isocyanate process have been
solved, at a cost, by engineering. Figure. 13.6 shows how a large core shop
collects and disposes of amine gas. Smaller foundries find the problems of
handling amine gas difficult so there is a need for a more environmentally
friendly process. Up to the present, only the CO_2 hardened ECOLOTEC
system meets the requirements.

A growing concern of all users of chemical binders is that of disposing of
used sand. All the developed countries impose severe restrictions on the
materials that can be safely dumped on landfill sites. Pollution of water
supplies by phenol and other leachable chemicals is a growing concern, so
all resin manufacturers are reducing the free-phenol content of their
products. The use of sand reclamation, both mechanical and thermal, is
being increasingly used by foundries to reduce the quantity of sand that
must be disposed to waste sites.

Chapter 14

Sodium silicate bonded sand

Introduction

The principles of the use, testing and reclamation of chemically hardened sand, including sodium silicate bonded sand, are described in the Introduction to Chapter 13.

Sodium silicate

Sodium silicate is a water soluble glass available from suppliers in a wide range of types specified by the silica (SiO_2), soda (Na_2O) and water content. Manufacturer's data sheets specify the "weight ratio" of silica to soda, the percentage of water and the viscosity. For foundry use, sodium silicates with ratios between 2 and 3 and water content around 56% are usually used.

Sodium silicate can be hardened in a number of ways: by adding weak acids (CO_2 gas or organic esters), by adding various powders (di-calcium silicate, anhydrite etc.) or by removing water. CO_2 gas and liquid ester hardeners are the most widely used of the silicate processes. Silicate binders have no smell and few health hazards, but the bond strength is not as high as that of resin binders and, being an inorganic bond, it does not burn out with heat, so breakdown after casting can be a problem. For this reason, various organic breakdown additives are often incorporated with the liquid silicate or added during sand mixing.

Typical data for foundry grade sodium silicate

	Wt. ratio	Na_2O (%)	SiO_2 (%)	H_2O (%)	s.g.	Visc.cP at 20°C	Litres/ tonne
Low ratio	2.00	15.2	30.4	54.4	1.56	850	641
Medium ratio	2.40	12.7	30.8	56.5	1.50	310	668
High ratio	2.85	11.2	31.9	56.9	1.48	500	677

CO$_2$ silicate process (basic process)

Principle: Sand is mixed with sodium silicate and the mixture blown or hand-rammed into a core box or around a pattern. Carbon dioxide gas is passed through the compacted sand to harden the binder.

Sand: Silica sand of AFS 50–60 is usually used, the process is quite tolerant of impurities and alkaline sand such as olivine can be used. Sand should be clay free. Low temperature retards hardening; the sand temperature should be above 15°C.

Binder addition: About 3.0–3.5% silicate addition is usually used. High ratio silicates give short gassing times, low gas consumption and improved post-casting breakdown, but high ratio silicate is easily over-gassed, resulting in reduced core strength and poor core storage properties. Low ratio binders give better core storage, but gassing times are longer and breakdown is not as good. Low ratio silicates (2.0–2.2) are usually used.

Mixing: Batch or continuous mixers are used; avoid excessive mixing time which may heat the sand and cause loss of water.

Bench life: Several hours if drying is prevented by storing in a closed container or covering with a damp cloth.

Core boxes: Any material; wood, resin or metal may be used but the boxes must be painted to prevent sticking. Polyurethane or alkyd paints are used followed by the application of a wax or silicone-wax polish, or STRIPCOTE AL.

Blowing: High pressure is needed, 650–700 kPa (90–100 psi).

Curing: The correct volume of gas is needed. Undergassing does not achieve optimum strength. Overgassing leads to reduction of strength on standing. To ensure correct hardening, a flow controller and timer should be used. If cores are to be stored for some time before use, it is preferable to undergas. Higher strength will then be developed on standing. Gassing times are typically 10–60 seconds, depending on the size of the core.

CO$_2$ gas consumption per kg of sodium silicate (good quality silica sand AFS 50)

Silicate ratio	Time to reach 700 kPa (100 psi) compression strength at 5 l/min	litres CO$_2$	kg CO$_2$
2.0	52 sec.	300	0.60
2.2	48	275	0.55
2.4	26	150	0.30
2.5	23	130	0.26

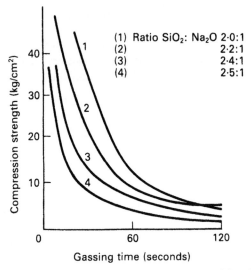

Figure 14.1   *The effect of 24 hour storage of cores gassed for different times as a function of silicate ratio.*

Core strength:

	as-gassed		after 6 hrs	
Tensile	350 kPa	50 psi	1000 kPa	150 psi
Compressive	2100	300	4200	600

Core storage: Overgassing causes cores to become friable during storage over 24 hours. Careful control of gassing is essential, Fig. 14.1

Casting characteristics: Cores require coating to produce good surface finish on ferrous castings. Spirit-based coatings should be used because water can soften the silicate bond. No metallurgical problems occur with ferrous or non-ferrous castings; the binder contains no nitrogen or sulphur. Breakdown after casting is poor (with ferrous castings) because silicate fuses with sand above 400°C.

Reclamation: Reclamation is difficult because the binder does not burn out during casting.

Health hazards: There is very little fume or smell during core manufacture, storage or casting. Gloves should be worn when handling silicates or mixed sand.

General comments: The CO_2 process was the first of the gas hardened processes to be introduced into the foundry and is still one of the easiest and cleanest processes to use. Foseco and other suppliers have made important improvements to the process while still retaining its ease of use and excellent environmental characteristics.

Gassing CO_2 cores and moulds

The core boxes are prepared by coating with a suitable parting agent; silicone waxes can be used but the best results are obtained with STRIPCOTE AL, an aluminised self-drying liquid.

The simplest gassing techniques are based either on a simple tube probe fitted through a backing plate, or a hood sealed to the open side of the box with a rubber strip, Fig. 14.2. Gas at a pressure of about 70–140 kPa (10–20 psi) is passed until the sand is felt to have hardened or some white crystals appear (indicating over–gassing). Venting of the boxes and patterns to prevent dead spots is essential. It is usual to rap the box before gassing to make stripping easier.

Control of gassing is improved by incorporating gas heaters (to prevent freezing of the valves), flow meters and timers so that automatic control of the volume of gas passed is possible. Not only does this practice economise on expensive CO_2, but the best strength and core storage life will only be achieved if the correct amount of gas is used. It is important that over gassing is avoided, any error should always be towards under-gassing. The suppliers of CO_2 gas will advise on the best equipment to use.

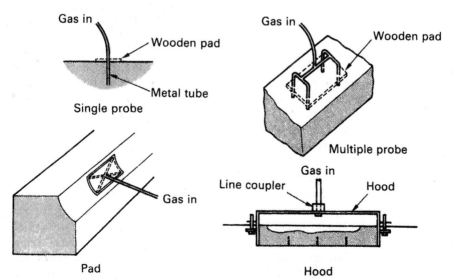

Figure 14.2 *CO_2-silicate process, methods of gas application.*

Improvements to the CO_2 silicate process

Foseco products: CARSIL ⎤   sodium silicate blended with special
                     SOLOSIL ⎦   additions
                     DEXIL   breakdown agent

Principle: The main drawbacks of the basic CO_2 process are:

poor breakdown of the bond after casting
poor core storage properties
rather low tensile strength

These properties can be greatly improved by special additives, while retaining the simplicity and user friendliness of the CO_2 process.

The CARSIL and SOLOSIL range: These products are a range of sodium silicate-based binders for the CO_2 process. They may be simple sodium silicates which can be used with DEXIL breakdown agent if required, or they may be "one-shot" products which incorporate a breakdown agent, or they may (like SOLOSIL) incorporate special additives to improve bond strength as well as breakdown.

Binders containing high levels of breakdown additives give improved post-casting breakdown but the maximum as-gassed strength is reduced and core storage properties are likely to be impaired. The selection of an optimum binder for a given application is therefore almost always a compromise. The requirement for high production rates and high as-gassed strength must be balanced against core storage properties and the need for good breakdown. The range of binders includes some which are suitable only for the CO_2 process, some which are suitable for self-setting applications and some which can be used for both processes.

The commonly used breakdown agents are organic materials which burn out under the effect of the heat of the casting. While solid breakdown agents such as dextrose monohydrate, wood flour, coal dust and graphite can be used, powder materials are not easy to add consistently to sand in a continuous mixer. Liquid breakdown agents are easier to handle, they usually consist of soluble carbohydrates. The best improve gassing speed without loss of strength. Some are also resistant to moisture pick-up and their use has increased the storage life of high ratio silicate bonded cores.

Sucrose is the only common carbohydrate soluble in sodium silicate without a chemical reaction. It is readily soluble up to 25% and many sugar- or molasses-based binders are available. Use of sucrose increases gassing speed but reduces maximum strength and storage properties. Nevertheless silicates containing sugar are the most popular CO_2 binders because of the convenience of a binder in the form of a single liquid. Molasses can be used as a low cost alternative to sugar, but it is subject to fermentation on storage. The Foseco CARSIL range of silicate binders is based on sugar. Some are designed for use with CO_2, others for self-setting (SS) with ester hardeners. Some can be used for both processes.

The CARSIL range of silicate binders

Product	Ratio	Additive	CO_2/SS	Comments
CARSIL 100	2.5:1	Sugar	CO_2/SS	Higher ratio for faster gassing, take care not to overgas. Can be used with ester hardeners.
CARSIL 513	2.4:1	Sugar	CO_2/SS	Low viscosity binder for easy mixing in continuous mixers. Moulds and cores.
CARSIL 520	2.0:1	High sugar	CO_2	High breakdown, low viscosity.
CARSIL 540	2.2:1	Low sugar	CO_2	Suitable for moulds or cores.
CARSIL 567	2.2:1	High sugar	CO_2	High breakdown, good for Al casting.

Note: Some of the CARSIL binders were formerly known as GASBINDA binders in the UK.

The extent to which a core will break down after casting varies depending on the type of metal cast. Low temperature alloys such as aluminium do not inject enough heat into the sand to burn out the breakdown agent fully. Indeed, the low temperature heating may even strengthen the core. In such cases it is useful to add additional breakdown agents such as DEXIL.

DEXIL 34BNF is a powder additive developed for use with light alloys. It also acts as a binder extender so reducing the silicate requirement. The application rate is 0.5–1.5%. It should be added to the sand and pre-dispersed before adding the silicate.

DEXIL 60 is a pumpable organic liquid which is particularly suitable for use with continuous mixers.

SOLOSIL

SOLOSIL was developed to improve on the performance of silicates containing sugar-based additives. SOLOSIL is a complex one-shot sodium silicate binder for the CO_2 gassed process. It contains a high level of breakdown agent/co-binder and offers a combination of high strength and rapid gassing with good core storage properties and excellent post-casting breakdown.

The binder is best used with good quality silica sand. Addition levels of 3.0–4.5% are used depending on the application. To take full advantage of

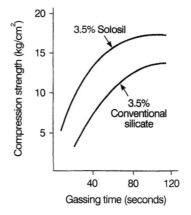

Figure 14.3   *Strength development of SOLOSIL compared with conventional sugar/silicate system.*

the high reactivity, an automatic gassing system incorporating a vaporiser, pressure regulator, flow controller and gassing timer is advisable. The high rate of strength development is shown in Fig. 14.3. While the transverse and tensile strength developed by SOLOSIL binders are still somewhat lower than some organic resin binders, SOLOSIL generally proves more cost effective and overcomes problems of poor hot strength, veining and finning, gas pinholing and fume on casting which occur with some resin binders.

Self-setting sodium silicate processes

The first self-setting process used powder hardeners. The Nishiyama process used finely ground ferrosilicon powder which reacts with sodium silicate generating heat and forming a very strong bond. The reaction also generates hydrogen which is dangerous. Other powder hardeners (which do not evolve dangerous gases) include di-calcium silicate, certain cements (such as blast furnace cement and sulphate resisting cement) and anhydrite. However, all powder hardeners are difficult to add uniformly to sand in continuous mixers, and their reactivity is difficult to control, since particle size and the age after grinding affect the reactivity of the powder. When liquid hardeners based on organic esters were introduced, the use of powder hardeners was largely discontinued.

Ester silicate process

Foseco products: CARSIL   sodium silicate binders
                     CARSET   ester hardener
                     VELOSET   special ester for very rapid setting

Principle: Sand is mixed with a suitable grade of sodium silicate, often incorporating a breakdown agent, together with 10–12% (based on silicate) of liquid organic ester hardener. The acid ester reacts with and gels the sodium silicate, hardening the sand. The speed of hardening is controlled by the type of ester used.

Sand: Dry silica sand of AFS 45–60 is usually used. As with all silicate processes, the quality and purity of the sand is not critical; alkaline sand such as olivine can be used. Fines should be at a low level. Sand temperature should be above 15°C; low temperature slows the hardening.

Additions: Sodium silicates with ratios between 2.2 and 2.8 are suitable, the higher the ratio, the faster the set. Silicates containing breakdown agents are usually used, additions between 2.5 and 3.5% are used depending on the sand grade. The ester hardener is commonly:

glycerol diacetate	fast cure
ethylene glycol diacetate	medium cure
glycerol triacetate	slow cure

Proprietary hardeners may be blends of the above with other esters. The addition level is 10–12% of the silicate.

Pattern equipment: Wood, resin or metal patterns can be used. Core boxes and patterns should be coated with polyurethane or alkyd paint followed by application of wax polish. STRIPCOTE parting agent may also be used.

Mixing: Continuous mixers are usually used; if batch mixers are used, the ester hardener should be mixed with the sand before adding the silicate.

Speed of strip: 20–120 minutes is common with normal ester hardeners. Attempts to achieve faster setting may result in lower strength moulds because the work time becomes short. With certain esters there is a tendency for core and mould distortion due to sagging if stripping occurs too early. Faster setting can be achieved by using the special VELOSET hardener.

Strength: The final strength achieved is:

Tensile	700 kPa (100 psi)
Compression	2000–5000 kPa (300–700 psi)

Coatings: Spirit-based coatings should be used.

Casting characteristics: No metallurgical problems arise with ferrous or non-ferrous castings. Breakdown is poor unless a silicate incorporating a breakdown agent is used.

Reclamation: As with all silicate processes, burnout of the bond does not occur during casting and attrition does not remove all the silicate residue so that build-up occurs in the reclaimed sand, reducing refractoriness and leading to loss of control of work time and hardening speed. The VELOSET system has been specially developed to permit reclamation (see below).

Environment: Silicate and ester have little smell and evolve little fume on casting. Silicates are caustic so skin and eye protection is needed while handling mixed sand.

CARSET 500 Hardeners: These are blends of organic esters formulated to give a wide range of setting speeds when used with sodium silicates, particularly the GARSIL series of silicates which incorporate a breakdown agent. For the best results, the silicate addition should be kept as low as possible in relation to the sand quality and the CARSET hardener maintained at 10% by weight of the silicate level. The speed of set is dependent on the sand temperature, silicate ratio and grade of CARSET hardener used.

The CARSET 500 series of hardeners

	Gel times (minutes) at 20°C using various CARSIL binders		
CARSET 500 series	*CARSIL 540 2.2 ratio*	*CARSIL 513 2.4 ratio*	*CARSIL 100 2.5 ratio*
500	8	7	5
511	9	8	6
522	13	12	8
533	19	15	9
544	105	53	21
555	–	–	90

Note: The gel time is the time taken for gelling to occur when silicate liquid is mixed with an appropriate amount of setting agent. The setting times may not be repeated exactly when sand is present, due to the possibility of impurities, but the figure provides a useful guide.

VELOSET hardeners: The VELOSET range is a series of advanced ester hardeners for the self-setting silicate process. They have been designed to give very rapid setting speed with a high strength, excellent through-cure and a high resistance to sagging. Used in the VELOSET Sand Reclamation Process, they provide the only ester silicate process in which the sand can be reclaimed by a simple dry attrition process and reused at high levels equal to those typical of resin bonded sands.

Additions: There are three grades of VELOSET hardener. VELOSET 1, 2 and 3. Binders of ratio 2.2–2.6 are used; lower ratios give inferior strength while if higher ratios are used the bench life becomes too short. The bench life obtained is independent of addition level. The level is usually 10–12% based on the binder. If the sand is to be reclaimed, the addition level of 11% should not be exceeded.

Bench life (minutes) at 20°C

CARSIL ratio	VELOSET grade		
	1	2	3
2.2	10	7	4
2.4	7	4	2
2.6	4	2	1

When a choice is possible, always use the highest ratio CARSIL binder and the slowest grade of VELOSET hardener. This provides optimum strength development.

Mixer: Since VELOSET is rapid setting, it is preferable to use a continuous mixer.

VELOSET sand reclamation process: With the conventional ester silicate process, dry attrition reclamation has occasionally been practised but the level of sand reuse is rarely more than 50%, which hardly justifies the capital investment involved. With the VELOSET system, up to 90% reuse of sand is possible using mechanical attrition.
    The process stages are:

    Crushing the sand to grain size
    Drying
    Attrition
    Classification
    Cooling

The reclaimed sand is blended with new sand in the proportion 75 to 25. During the first 10 cycles of reuse, the sand system stabilises and the bench life of the sand increases by a factor of up to 2. Also, mould strength should improve, and it is usually possible to reduce the binder addition level by up to 20% yet still retaining the same strength as achieved using new sand. Once the process has become established, it may become possible to reuse up to 85–90% of the sand, Figs 14.4 and 14.5.

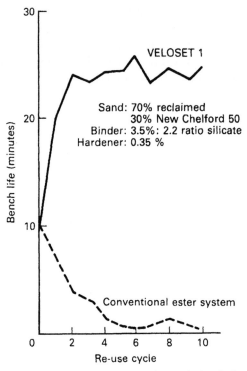

Figure 14.4   *VELOSET reclamation, showing the variation in bench life after repeated use of relaimed sand, compared with conventional ester process.*

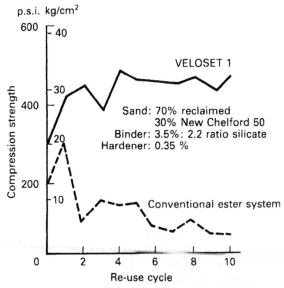

Figure 14.5   *VELOSET reclamation, ultimate strength characteristics of reclaimed sand, compared with conventional ester process.*

Adhesives and sealants

It is often necessary to joint cores together to form assemblies, or to glue cores to moulds before closing the mould. A range of CORFIX adhesives is available:

CORFIX grade	Type	Set time	Temp(°C)	Remarks
4	Stove hardening	30	180–220	High viscosity gap filling
8	Air hardening	slow	ambient	For CO_2 and self-set silicate
21	Air hardening	fast	ambient	Any cold core
25	Hot melt	open time 15–120 sec.	140–180	Core assembly at high rates, shell process

CORSEAL sealants

This is a group of core sealing or mudding compounds for filling out joint lines, cracks and minor blemishes in cores. CORSEAL is available in two forms:

CORSEAL 2 is a powder which is mixed with water to form a thick paste (4 parts product to 3 parts water). The paste is applied by spatula or trowel (or fingers) and allowed to dry for about an hour. It may be lightly torched if required immediately.

CORSEAL 3 and 4 are ready-mixed self-drying putties which are sufficiently permeable when full dry to prevent blowing but strong enough to prevent metal penetration into the joint. Drying time depends on local conditions and the thickness of the layer applied but should be at least 30 minutes.

TAK sealant

Small variations in the mating faces of moulds due to flexing of patterns or deformation of moulding boxes and moulding materials may result in gaps into which liquid metal will penetrate causing runout and flash. This can be prevented by the application of TAK plastic mould sealant which forms a metal and gas-tight seal. TAK does not melt at high temperatures and, if metal touches it, it burns to a compact, fibrous mass. The TAK strip is laid around the upper surface of the drag mould, about 25 mm from the edge of the mould cavity and the mould is then closed and clamped. TAK can also be used to seal small core prints:

TAK 3 is supplied in cartridge form for extrusion from a hand gun; a variety of nozzle sizes is available.

TAK 500 is ready-extruded material supplied in continuous lengths of 6 mm diameter.

Chapter 15
Magnesium casting

Casting alloys

Magnesium alloy castings are used for aerospace, automotive and electronic applications. Their main advantage is their light weight; typical magnesium alloys have a density of 1.8 g/ml compared with 2.7 g/ml for aluminium alloys. Aluminium is the principal alloying constituent of magnesium-based casting alloys with zinc and manganese also present in small amounts. Pressure diecasting is the most commonly used casting process and because of the low casting temperature (650–700°C), hot chamber diecasting machines can be used. Magnesium diecastings can be made with thinner walls than aluminium, allowing the overall weight of components to be substantially reduced and compensating for the higher alloy cost per kilogram. Gravity diecasting and sand casting are also used, particularly for more highly stressed castings. The use of high purity alloys with low levels of Fe, Ni and Cu improves corrosion resistance allowing their use in automotive applications exposed to road salt. The use of magnesium alloy diecastings in automotive components is growing rapidly as automobile companies seek ways of reducing weight. Some vehicles already contain as much as 10–20 kg of Mg components. The most popular parts made at present for production cars are: instrument panel substrates, cross car

Table 15.1 Commonly used magnesium alloys

Alloy	Characteristics	Typical uses
AZ91 AZ81	The most common alloys for pressure and gravity die and sand casting	Housings, covers, brackets, chain saw parts, hand tools, computer parts etc.
AM50 AM60	Both alloys combine strength, ductility castability and cold workability	Seat frames, instrument panels, brackets, wheels.
AM20	Used for pressure diecastings where high ductility and impact strength are required	Automotive safety parts.

Table 15.2   Composition of magnesium alloys

				Composition				
Alloy	*Al*	*Zn*	*Mn*	*Cu*	*Fe*	*Si*	*Ni*	*Total impurities*
AZ91	8.0–9.5	0.3–1.0	0.1–0.3	0.15	0.05	0.3	0.01	0.40
AZ81	7.5–9.0	0.3–1.0	0.15–0.4	0.15	0.05	0.3	0.01	0.40
AM50	4.5–5.3	0.1	0.27 min	0.008	0.004	0.1	0.001	
AM60	5.7–6.3	0.2	0.27 min	0.008	0.004	0.05	0.001	
AM20	1.7–2.2	0.1	0.5 min	0.008	0.004	0.1	0.001	

Note: Single figures are maximum %.
High purity versions of AZ91 and AZ81 are frequently used, they have max. Fe 0.004, Ni 0.001, Cu 0.015, Si 0.05, others 0.01 each.
The above figures are intended as a guide only. National specifications may differ and must be referred to.
Mechanical properties are similar to the commonly used aluminium alloys, Table 15.3.

beams, seat frames. Wheels, gearbox casings, sumps and inlet manifolds are used on Formula One and other racing cars.

The most commonly used casting alloys (using the ASTM designation, which is frequently used) are described in Tables 15.1, 15.2 and 15.3. Mg–zirconium and Mg–yttrium high strength alloys have been developed and are used mainly for defence applications.

The melting, treatment and casting of magnesium alloys

Molten magnesium alloys attack firebrick and refractory furnace linings resulting in harmful silicon contamination. For this reason, steel crucibles, pressed or cast, are used. Iron is also slightly soluble in magnesium alloys but it has a much less harmful effect than silicon. Scrap should be cleaned and if possible shot-blasted to remove adhering sand as a further precaution against silicon pick-up. To eliminate ladling, the molten alloy should, if possible, be poured direct from the melting pot.

Magnesium alloys must be melted under covering and cleansing fluxes, to avoid oxidation losses and to remove inclusions. Inhibitor powders should be used to cover exposed metal during holding and pouring, and added to moulding sand to prevent chemical reaction. Magnesium alloys benefit from grain refinement which is carried out by inoculating with carbonaceous materials. Hexachloroethane is effective, decomposing in the liquid metal to

Table 15.3 Mechanical properties of magnesium alloys

			Typical mechanical properties				
Alloy	Form	Condition	TS (N/mm²)	YS (N/mm²)	Elong. (%)	Brinell hardness	Melting range (°C)
AZ91HP	Press. die		200–250	150–170	0.5–3.0	65–85	420–600
	Grav. die	F	160–220	110–130	2–5	55–70	
		T4	240–280	120–160	6–10	55–60	
		T6	240–300	150–190	2–7	60–90	
	Sand	F	160–220	90–120	2–5	50–65	
		T4	240–280	110–140	6–12	55–70	
		T6	240–300	150–190	2–7	60–90	
AZ81HP	Press. die		200–240	140–160	1–3	60–85	425–615
	Grav. die	F	160–220	90–110	2–6	50–65	
		T4	240–280	90–120	8–12	50–65	
	Sand	F	160–220	90–110	2–6	50–65	
		T4	240–280	90–120	8–12	50–65	
AM60HP	Press. die		190–230	120–150	4–8	55–70	445–630
	Sand	F	180–240	80–110	8–12	50–65	
		T4	190–250	90–110	8–15	50–65	
AM50HP	Press. die		180–220	110–140	5–9	50–70	440–625
AM20HP	Press. die		160–210	90–120	8–12	40–55	
Aluminium alloys for comparison							
LM24	Al–Si8Cu3Fe						
	Press. die	F	110	200	2	85	
M25	Al–Si7Mg						
	Grav. die	F	180	90	5	60	
	Sand	F	140	90	2.5	60	
		TE	170	130	1.5	70	

Notes: F = as cast
      T4 = solution treated
      T6 = solution treated and artificially aged
      TE = precipitation treated (Al alloy).
      The data is intended as a guide only, refer to National Standards for details.

form specks of carbon throughout the melt which act as nuclei for grain growth. The use of hexachloroethane in aluminium alloy metal treatment has been banned in Europe for health and safety reasons, although it is still permitted for grain refining magnesium until alternative treatments have been developed. Foseco has withdrawn all hexa-containing products from sale.

Table 15.4   Magnesium–Zirconium alloys

Alloy (ASTM)	Zn	RE metals	Zr	Cu	Ni
ZK51	3.5–5.5	–	0.4–1.0	0.03	0.005
ZE41	3.5–5.5	0.75–1.75	0.4–1.0	0.03	0.005
EZ33	0.8–3.0	2.5–4.0	0.4–1.0	0.03	0.005

Note: Single figures are maximum %.

Mechanical properties:

Alloy	Form	Tensile strength (MPa)	Elongation (%)	Properties
ZK51	Sand cast	230	5	High strength, good ductility
	Chill cast	245	7	
ZE41	Sand cast	200	3	High strength, pressure-tight
	Chill cast	215	4	
EZ33	Sand cast	140	3	Pressure-tight at high
	Chill cast	155	3	temperature

Table 15.5   Magnesium–yttrium alloys

Alloy (ASTM)	Zn	RE*	Zr	Cu	Ni	Fe	Si	Mn	Yt	TS (MPa)	Elong.
WE54	0.2	2.0–4.0	0.4–1.0	0.03	0.005	0.01	0.01	0.15	4.75–5.5	250	2%
WE43	0.2	2.4–4.4	0.4–1.0	0.03	0.005	0.01	0.01	0.15	3.7–4.3	250	2%

*Neodynium and heavy rare earths.
WE54 has good strength up to 300°C for short times.
WE43 has good strength up to 250°C for long times.

Melting

MAGREX 60 flux is used as a covering and cleansing flux; it provides a liquid surface cover during melting which prevents contact with the air so that melting losses are reduced and burning prevented. It also has a scavenging action which removes non-metallic impurities.

A little MAGREX 60 is dusted into the bottom of a heated, pressed steel crucible. The ingot and scrap are then charged on top and a further addition of MAGREX made. The total application should be approximately 1% of the charge weight. Melt down rapidly, maintaining a good cover at all times.

At about 750°C the heat should be stopped, the crucible sides scraped and the melt skimmed. A further 2% addition of MAGREX is then made and rabbled in well with a perforated plunger. More MAGREX is added progressively and stirred until the metal surface, which previously had a frothy appearance, becomes bright. As the MAGREX absorbs oxides and impurities, its density increases until it sinks to the bottom of the crucible.

During the time the melt is cooling to the correct temperature, it should be skimmed and the cleaned area immediately dusted with an inhibitor powder, such as "flowers of sulphur", to prevent burning. Pour carefully, dusting the metal stream as it enters the mould with sulphur to prevent oxidation. Care must be taken to keep back any slag and particularly when nearing the end of the pour, to prevent any sludge entering the mould. Remove the sludge from the bottom of the pot and thoroughly scrape the sides and bottom before returning it to the furnace for recharging.

Use of sulphur hexafluoride

Fluxless melting of Mg alloys requires another form of melt protection. Sulphur hexafluoride, SF_6, is a colourless, odourless gas having low toxicity. At low concentrations, for example less than 0.8% in air or air/CO_2, it promotes the formation of a protective film on liquid magnesium which prevents oxidation. SF_6, like other fluorine-containing gases, is a "green-

Table 15.6   Use of sulphur hexafluoride in pressure diecasting operations

			Operating conditions	
Melt temp. (°C)	Recommended atmosphere over the melt (vol %)	Surface agitation	Residual flux**	Melt protection
650–705	air + 0.04 SF_6*	No	No	Excellent
650–705	air + 0.2 SF_6	Yes	No	Excellent
650–705	75 air + 25 CO_2 + 0.2 SF_6	Yes	Yes	Excellent
705–760	50 air + 50 CO_2 + 0.3 SF_6	Yes	No	Excellent
705–760	50 air + 50 CO_2 + 0.3 SF_6	Yes	Yes	Very good

*Minimum concentration under controlled conditions.
**May be present from prior operations.
Note: High humidity either in the outer atmosphere surrounding the melt or in the air blended with the SF_6/CO_2 will reduce the effectiveness of SF_6. Dry air (less than 0.1% H_2O by volume) should be used in the mixing.

Table 15.7   Use of sulphur hexafluoride in gravity casting operations (up to 830°C)

Crucible diameter	Quiescent (melting/holding) low gas flow rate		Agitated (alloying/pouring) high gas flow rate	
	SF_6 (ml/min.)	CO_2 (l/min.)	SF_6 (ml/min.)	CO_2 (l/min.)
30 cm	60	3.5	200	10
50 cm	60	3.5	550	30
75 cm	90	5	900	50

Note: These suggested flow rates are 1.7%–2% SF_6 by volume. The use of SF_6/CO_2 atmospheres for melting yttrium-containing alloys can lead to yttrium loss by preferential oxidation by CO_2. Argon/SF_6 atmospheres are recommended for these alloys during melting and holding.

house gas" considered harmful to the atmosphere and its use must be minimised. The most common gases used in fluxless melting are SF_6 mixed with dry air and some CO_2. The recommended protective atmospheres under various operating conditions are shown in Tables 15.6 and 15.7 taken from data provided by the International Magnesium Association.

The concentrations in Tables 15.6 and 15.7 should be maintained close to the melt surface. A gas mixing unit is recommended to control both the flow rate and the concentration of the gas. The protective atmosphere should be supplied through a manifold with several outlet nozzles positioned to supply gas to the whole surface of the melt. The furnace cover design is important for conservation of SF_6.

Casting temperature

Light castings, under 15 mm	780–810°C
Medium castings, 15–40 mm	760°C
Heavy castings, over 40 mm	730°C

Use of inhibitors in moulding sand

A chemical inhibitor must be added to moulding sand to prevent reaction between the molten magnesium and the moisture present in the green sand. Several materials can be used including sulphur, boric acid or ammonium bifluoride, used singly or together. The amount needed varies, depending on the moisture content of the sand and section thickness of the casting. Generally, however, 4–6% sulphur with 0.5% boric acid is used or up to 2% of ammonium bifluoride alone.

Recovering and refining magnesium

MAGREX 60 flux is suitable for the melting and refining of magnesium alloy foundry returns and turnings. 5–10% of MAGREX 60 is prefused in the melting unit and the dry scrap gradually added through the flux. After all the additions have been made and the melt is at 700–750°C, a further 2% of flux is fused on the melt before rabbling well into the metal for 3 to 5 minutes. The cleansed metal is then decanted off the spent flux. After pouring, the spent flux and impurities are scraped from the sides and bottom of the melting unit.

Foundry tools can be cleansed by immersing them in molten MAGREX 60 flux for a few minutes. This will absorb any adhering metal oxides and leave them clean and ready for use.

Running and gating

Since magnesium alloys oxidise rapidly, every effort should be made to ensure non-turbulent pouring. Use of SIVEX FC or STELEX ceramic foam filters to remove oxide films is recommended (see Chapter 8). As magnesium is so light, a pouring basin should be built up above the top surface of the mould to provide greater metallostatic pressure. Running and feeding methods should be as for a moderately high shrinkage alloy, e.g. fairly large diameter runners and large feeding heads with KALMIN insulating sleeves.

Gravity diecasting

Magnesium alloys are so light in weight that little metallostatic pressure is available to displace mould air and the possibility of short runs or cold shuts is enhanced. Dies should be designed with ample venting and down-sprues should be large in area relative to aluminium practice. The minimum wall thickness of castings should be 5 mm. Die coatings are the same as for aluminium alloys.

The alloy for diecasting is frequently melted in a fully enclosed bale-out furnace under an inert atmosphere of sulphur dioxide or sulphur hexa-fluoride gas mixtures (Table 15.7). MAGREX 60 is used. As it becomes impregnated with oxides, it sinks to the bottom of the melt. When the metal bath is at a temperature of 720–750°C, about 3% of MAGREX is introduced. As the charge is replenished from time to time, more MAGREX is added. The flux cover is stirred from 3 to 4 minutes until the alloy surface is bright. The spent flux must be removed from the bottom of the furnace daily.

Pressure diecasting

Mg alloys can be diecast using both cold chamber and hot chamber machines. Hot chamber machines are so called because they use a pump

Table 15.8   Estimated weight of diecast magnesium automotive components

Component	Estimated weight for a mid-size car (kg.)
Dashboards	3.0–5.0
Bumper holders	2.5–4.5
Holders and supports	1.0–2.0
Front seat frames	18.0–24.0
Electronic circuitry cases	0.2–0.7
Gearboxes	8.0–12.0
Cylinder head covers	0.5–1.2
Oil sumps	0.8–1.2
Pedal supports	1.0–1.8
Wheels	14.0–26.0
Steering wheels	0.3–0.5

submerged in the molten alloy to fill the die and apply the required pressure during solidification, the cold chamber machines require a measured quantity of molten metal to be transferred from a holding furnace to the machine for each shot. The hot chamber process can achieve higher production rates and the castings produced are generally more consistent.

Pressure diecasting is the most frequently used process for automotive castings, where the growth in usage is high because of the attraction of weight reduction. Potential magnesium components and their estimated weights are shown in Table 15.8 (reproduced by courtesy of Buhler Ltd).

The usage of magnesium diecastings is expected to increase from 51 000 tonnes in 1996 to 186 000 tonnes in the year 2006.

Chapter 16

Copper and copper alloy castings

The main copper alloys and their applications

1 High conductivity coppers. Used chiefly for their high electrical and thermal conductivities. Applications include tuyeres for blast furnaces and hot blast cupolas, water-cooled electrode clamps, switchgear etc.

2 Brasses; copper–zinc alloys where zinc is the major alloying element. Easy to cast, with excellent machinability and good resistance to corrosion in air and fresh water. They are widely used for plumbing fittings. High tensile brasses are more highly alloyed and find uses in marine engineering.

3 Tin bronzes; copper–tin alloys where tin is the major alloying element. With tin contents of 10–12%, tin bronze castings are more expensive than brass. They have high corrosion resistance and are suitable for handling acidic waters, boiler feed waters etc. High tin alloys are also used in wear-resistant applications.

4 Phosphor bronzes; copper–tin alloys with an addition of about 0.4–1.0% P. They are harder than tin bronzes but with lower ductility. They are used for bearings where loads and running speeds are high and for gears such as worm wheels.

5 Lead bronzes; copper–tin–lead alloys. Used almost exclusively for bearings, where loads and speeds are more moderate.

6 Gunmetals; copper–tin–zinc–lead alloys. Favourite alloys for sand casting. They have a good combination of castability, machinability and strength with good corrosion resistance. They are used for intricate, pressure-tight castings such as valves and pumps. Also for bearings where loads and speeds are moderate.

7 Aluminium bronzes; copper–aluminium alloys where Al is the major alloying element. They combine high strength with high resistance to corrosion. Applications range from decorative architectural features to highly stressed engineering components. They have many marine uses including propellers, pumps and valves and are used for the manufacture of non-sparking tools.

8 Copper–nickels; copper–nickel alloys where Ni is the major alloying element. Used for marine applications in severe conditions, for example for pipework.

(The above information is based on data kindly supplied by the Copper Development Association, St Albans, Herts.)

Specifications for copper-based alloys

The new BS EN 1982 standard for Copper and Copper Alloy Ingots and Castings is the British implementation of the European Standard and it replaces BS 1400:1985 which has been withdrawn. BS 1400 used abbreviations of the type of material:

SCB	sand casting brass
DCB	diecasting brass
HTB	high tensile brass
DZR	dezincification resistant brass
HCC	high conductivity copper
CC	copper–chromium
CT	copper–tin (bronze)
PB	phosphor bronze
LB	leaded bronze
LG	leaded gunmetal
G	gunmetal
AB	aluminium bronze
CMA	copper–manganese–aluminium
CN	copper–nickel

These have been superseded in the European Standard by compositional designations with the base metal first followed by the major alloying elements, e.g. CuZn33Pb2-C is a leaded brass casting alloy containing 33% Zn and 2% Pb. Two further letters are used to designate the relevant casting process which affects the mechanical properties:

GM	permanent mould casting
GS	sand casting
GZ	centrifugal casting
GP	pressure diecasting
GC	continuous casting

Note that there is not necessarily an exact equivalence between the BS 1400 alloy and the corresponding BS EN alloy. The European Standard also uses a Material Designation Number for each casting alloy so the leaded brass referred to above is designated:

> CuZn33Pb2-C Number CC750S and is equivalent to the old BS 1400 sand casting brass SCB3.

Table 16.1 lists the BS EN 1982 alloys and the nearest BS1400 equivalent alloys which they replace.

Table 16.2 lists the compositions and mechanical properties of the alloys.

Thanks are due to the Copper Development Association (Verulam Industrial Estate, 224 London Road, St Albans, Herts AL1 1AQ, England) for providing the information in Tables 16.1 and 16.2.

Colour code for ingots

In the UK, the following system was used for colour coding of ingots.

Designation Group A	Colour code	Designation Group B	Colour code
PB4	Black/red	PB1	Yellow
LPB1	Black	PB2	Yellow/red
LB2	White	CT1	Black/aluminium
LB4	White/green	LB5	White/brown
LG2	Blue	LG1	Blue/red
LG4	Blue/brown	AB1	Aluminium
SCB1	Green/blue	AB2	Aluminium/green
SCB3	Green	CMA1	Aluminium/red
SCB6	Green/brown	CMA2	Aluminium/yellow
DCB1	Yellow/blue	HTB1	Brown
DCB3	Yellow/brown	HTB3	Brown/red
PCB1	White/blue		
Group C			
LB1	White/black		
G3	Blue/black		
SCB4	Green/yellow		
G1	Red		

Note: Group A are alloys in common use.
Group B are special purpose alloys.
Group C are alloys in limited production.

Table 16.1 Copper and copper alloy ingots and castings – comparison of BS1400 and BS EN 1982 Showing near equivalents where standardised in BS EN 1982 and original compositional symbols for guidance where no near equivalent is included. See Table 16.2 for full details of composition and properties.

Nearest equivalent in old BS 1400 or BS4577	BS EN or ISO symbol for castings (1)	BS EN material designation number for castings (2)	BS EN relevant casting processes and designations (3)				
			GM Diecasting	GS Sand	GZ Centrifugal	GP Pressure-die	GC Continuous
Cooper and Copper-chromium (High conductivity coppers)							
HCC1	Cu-C	CC040A	✓	✓			
CC1–TF	CuCr1–C	CC140A	✓	✓			
A4/1	G–CuNiP						
A3/2	G–CuNi2Si						
A3/1	G–CuCo2Be						
A4/2	G–CuBe						
Copper–zinc (Brasses)							
DZR1	CuZn35Pb2Al–C	CC752S	✓			✓	
DZR2	CuZn33Pb2Si–C	CC751S	✓			✓	
–	CuZn37Pb2Ni1AlFe–C	CC753S					
PCB1	G–CuZn40Pb						
DCB1	CuZn38Al–C	CC767S	✓				
DCB2	G–CuZn37Sn						
DCB3	CuZn39Pb1Al–C	CC754S	✓		✓	✓	
–	CuZn39Pb1AlB–C	CC755S	✓			✓	
SCB1	G–CuZn25Pb35n2						
SCB2	G–CuZn30Pb3						
SCB3	CuZn33Pb2–C	CC750S		✓	✓		
SCB4	G–CuZn365n						
SCB5	G–CuZn105n						
SCB6	CuZn15As–C	CC760S	✓	✓			
–	CuZn16Si4–C	CC761S		✓			
–	CuZn32Al2Mn2Fe1–C	CC763S		✓	✓		
–	CuZn34Mn3Al2Fe1–C	CC764S	✓	✓	✓		✓
HTB1	CuZn35Mn2Al1Fe1–C	CC765S	✓	✓	✓		✓
HTB2	G–CuZn36Al4FeMn						
HTB3	CuZn25Al5Mn4Fe3–C	CC762S	✓	✓	✓		
–	CuZn37Al1–C	CC766S	✓	✓	✓		
Copper–tin (Gunmetals and Phosphor-bronzes)							
CT1	CuSn10–C	CC480K	✓	✓	✓		✓
PB1	CuSn11P–C	CC481K	✓	✓	✓		✓
–	CuSn11Pb2–C	CC482K		✓	✓		✓

Designation	Composition	Number	GM	GS	GZ	GP	GC
PB2	CuSn12–C	CC483K	✓	✓	✓		✓
CT2	CuSn12Ni2–C	CC484K	✓	✓	✓		✓
PB4	G–CuSn10PbP						
LPB1	G–CuSn7PbP						
Copper–tin–lead (Gunmetals and Leaded bronzes)							
LG1	CuSn3Zn8Pb5–C	CC490K		✓	✓		✓
LG2	CuSn5Zn5Pb5–C	CC491K	✓	✓	✓		✓
LG3	G–CuSn7Pb4Zn2						
LG4	CuSn7Zn2Pb3–C	CC492K	✓	✓	✓		✓
–	CuSn7Zn4Pb7–C	CC493K	✓	✓	✓		✓
LB1	CuSn7Pb15–C	CC496K		✓	✓		✓
LB2	CuSn10Pb10–C	CC495K	✓	✓	✓		✓
LB3	G–CuSn10Pb5						
LB4	CuSn5Pb9–C	CC494K	✓	✓	✓		✓
LB5	CuSn5Pb20–C	CC497K		✓	✓		✓
G1	G–CuSn10Zn2						
G2	G–CuSn8Zn4Pb						
G3	G–CuSn7Ni5Zn3						
Copper–aluminium (Aluminium bronzes)							
–	CuAl9–C	CC330G	✓	✓	✓		
AB1	CuAl10Fe2–C	CC331G	✓	✓	✓		
–	CuAl10Ni3Fe2–C	CC332G	✓	✓	✓		
AB2	CuAl10Fe5Ni5–C	CC333G	✓	✓	✓		✓
–	CuAl11Fe6Ni6–C	CC334G	✓	✓	✓		✓
AB3	G–CuAl6Si2Fe		✓	✓	✓		✓
Copper–manganese–aluminium							
CMA1	CuMn11Al8Fe3Ni3–C	CC212E		✓	✓		
CMA2	G–CuMn13Al9Fe3Ni3						
Copper–nickel (cupro-nickels)							
–	CuNi10Fe1Mn1–C	CC380H		✓	✓		
–	CuNi30Fe1Mn1–C	CC381H		✓	✓		
CN1	CuNi30Cr2FeMnSi–C	CC382H		✓			
CN2	CuNi30Fe1Mn1NbSi–C	CC383H		✓			✓

(1) Symbol finishes with B for material in ingot form.

(2) Number begins CB for material in ingot form.

(3) GM – permanent mould casting. GS – sand casting. GZ – centrifugal. GP – pressure diecasting. GC – continuous casting. Method of casting affects properties significantly.

Note: Ingots are not specified for high conductivity coppers.

Table 16.2   BS EN 1982 copper and copper alloy ingots and castings–compositions, uses and typical propertie

Material designation		Composition, %, range or max									
Symbol for castings (1)	Number for castings (2)	Cu	Al	Fe	Mn	Ni	P	Pb	Sn	Zn	Others
Copper and Copper–chromium (High conductivity coppers)											
Cu–C	CC040A	None specified									
CuCr1–C	CC140A	Rem.									0.4–1.2
Copper–zinc (Brasses)											
CuZn33Pb2–C	CC750S	63.0–66.0				1.0		1.0–3.0	1.5	Rem.	
CuZn33Pb2Si–C	CC751S	63.5–66.0		0.25–0.5		0.8		1.3–2.2		Rem.	0.065–
CuZn35Pb2Al–C	CC752S	61.5–64.5	0.3–0.70					1.5–2.5		Rem.	0.15 A
CuZn37Pb2Ni1AlFe–C	CC753S	58.0–61.0	0.4–0.8	0.5–0.8		0.5–1.2		1.8–2.50	0.8	Rem.	
CuZn39Pb1Al–C	CC754S	58.0 63.0	0.8			1.0		0.5–2.5	1.0	Rem.	
CuZn39Pb1AlB–C	CC755S	59.5–61.0		0.05–0.2				1.2–1.7		Rem.	(3)
CuZn15As–C	CC760S	83.0–88.0								Rem.	0.05–0
CuZn16Si4–C	CC761S	78.0–83.0	0.1			1.0		0.8		Rem.	3.0–5.0
CuZn25Al5Mn4Fe3–C	CC762S	60.0–67.0	3.0–7.0	1.5–4.0	2.5–5.0	3.0				Rem.	
CuZn32Al2Mn2Fe1–C	CC763S	59.0–67.0	1.0–2.5	0.5–2.0	1.0–3.5	2.5		1.5	1.0	Rem.	1.0 Si
CuZn34Mn3Al2Fe1–C	CC764S	55.0–66.0	1.0–3.0	0.5–2.5	1.0–4.0	3.0				Rem.	
CuZn35Mn2Al1Fe1–C	CC765S	57.0–65.0	0.5–2.5	0.5–3.0	6.0			1.0		Rem.	
CuZn37Al1–C	CC766S	60.0–64.0	0.3–1.8			2.0				Rem.	
CuZn38Al–C	CC767S	59.0–64.0	0.1–0.8			1.0				Rem.	
Copper–tin (Gunmetals and Phosphor-bronzes)											
CuSn10–C	CC480K	88.0–86.0				2.0	0.2	1.0	9.0–11.0		
CuSn11P–C	CC481K	87.0–89.5					0.5–1.0		10.0–11.5		
CuSn11Pb2–C	CC482K	83.5–87.0				2.0	0.40	1.0–2.5	10.5–12.5	2.0	
CuSn12–C	CC483K	85.0–88.5				2.0	0.60	0.7	11.0–13.0		
CuSn12Ni2–C	CC484K	84.5–87.5				1.5–2.5	0.05–0.40		11.0–13.0		
Copper–tin–lead (Gunmetals and Leaded bronzes)											
CuSn3Zn8Pb5–C	CC490K	81.0–86.0				2.0	0.05	3.0–6.0	2.0–3.5	7.0–9.5	
CuSn5Zn5Pb5–C	CC491K	83.0–87.0				2.0	0.01	4.0–6.0	4.0–6.0	4.0–6.0	
CuSn7Zn2Pb3–C	CC492K	85.0–89.0				2.0	0.01	2.5–3.5	6.0–8.0	1.5–3.5	
CuSn7Zn4Pb7–C	CC493K	81.0–85.0				2.0	0.10	5.0–8.0	6.0–8.0	2.0–5.0	
CuSn5Pb9–C	CC494K	80.0–87.0				2.0	0.10	8.0–10.0	4.0–6.0	2.0	
CuSn10Pb10–C	CC495K	78.0–82.0				2.0	0.10	8.0–11.0	9.0–11.0	2.0	
CuSn7Pb15–C	CC496K	74.0–80.0				0.5–2.0	0.10	13.0–17.0	6.0–8.0	2.0	
CuSn5Pb20–C	CC497K	70.0–78.0				0.5–2.0	0.10	18.0–23.0	4.0–6.0	2.0	
Copper–aluminium (Aluminium bronzes)											
CuAl9	CC330G	88.0–92.0	8.0–10.5	1.2	0.50	1.0					
CuAl10Fe2–C	CC331G	83.0–89.5	8.5–10.5	1.5–3.5	1.0	1.5					
CuAl10Ni3Fe2–C	CC332G	80.0–86.0	8.5–10.5	1.0–3.0	2.0	1.5–4.0					
CuAl10Fe5Ni5–C	CC333G	76.0–83.0	8.5–10.5	4.0–5.5	3.0	4.0–6.0					
CuAl11Fe6Ni6–C	CC334G	72.0–78.0	10.0–12.0	4.0–7.0	2.5	4.0–7.5					
Copper–manganese–aluminium											
CuMn11Al8Fe3Ni3–C	CC212E	68.0–77.0	7.0–9.0	2.0–4.0	8.0–15.0	1.5–4.5					
Copper–nickels											
CuNi10Fe1Mn1–C	CC380H	84.5		1.0–1.8	1.0–1.5	9.0–11.0					
CuNi30Fe1Mn1–C	CC381H	64.5		0.5–1.5	0.6–1.2	29.0–31.0					
CuNi30Cr2FeMnSi–C	CC382H	Rem.		0.5–1.0	0.5–1.0	29.0–32.0					0.15 Zr 1.5–2.0 0.25 Ti
CuNi30Fe1Mn1Ni0i–C	CC383H	Rem.		0.5–1.5	0.6–1.2	29.0–31.0					0.5–1.0

(1) Symbol finishes-B for material in ingot form.
(2) Number begins-CB for material in ingot form.
(3) Boron for grain refining.
(4) GM – permanent mould casting. GS – sand casting. GZ – centrifugal. GP – pressure diecasting. GC – continuous casting
Method of casting affects properties significantly.
Note: Ingots are not specified for high conductivity coppers.

Characteristics and uses	Typical minimum mechanical properties (properties vary significantly with method of casting)				Relevant casting processes and designations (4)				
					GM Diecasting	GS Sand	GZ Centrifugal	GP Pressure-die	GC Continuous
	0.2% proof strength (N/mm²)	Tensile strength (N/mm²)	Elongation (%)	Hardness (HB)					
Electrical and thermal applications. Additionally specified by minimum conductivity requirements	40	150	25	40	✓	✓			
	250	350	10	95	✓	✓			
General purpose applications	70	180	12	45		✓	✓		
Dezincification resistant alloys for water fittings in areas with aggressive waters	280	400	5	110				✓	
	150	300	10	90	✓		✓		
Fine grained, freely machinable	150	300	15	90	✓				
General purposes, extensively for plumbing fittings. Boron in CC755S gives superior strength for thin sections	120	280	10	70	✓	✓	✓	✓	
	180	350	10	90	✓			✓	
Brazable, Good corrosion resistance	70	160	20	45		✓			
Silicon brass for valves and water fittings	300	500	8	130	✓	✓		✓	
High tensile brasses for engineering castings when good wear resistance is required and high loads encountered	480	750	5	190	✓	✓	✓		✓
	200	430	8	110		✓		✓	
	260	600	12	140	✓	✓	✓		
	200	480	18	110	✓	✓	✓		✓
General purpose, high quality engineering castings	170	450	25	105	✓				
	130	380	30	75	✓				
For gears and general bearing applications offering higher corrosion/ erosion resistance than gunmetals. The lead in CC482K gives improved machinability and the nickel in CC484K increases strength and hardness	160	270	10	80	✓	✓	✓		✓
	170	310	4	85	✓	✓	✓		✓
	150	280	5	90		✓	✓		✓
	150	270	5	85	✓	✓	✓		✓
	180	300	10	95	✓	✓	✓		✓
Leaded gunmetals giving good corrosion resistance with moderate strength and good castability. Applications include valves and bearings	100	220	12	70		✓	✓		✓
	110	230	10	65	✓	✓	✓		✓
	130	240	12	70	✓	✓	✓		✓
	120	240	12	70	✓	✓	✓		✓
Leaded tin-bronzes whose plasticity increases with lead content for bearings when some measure of plasticity is required and for when there is a risk of scoring mating materials	80	200	6	60	✓	✓	✓		✓
	110	220	5	65	✓	✓	✓		✓
	90	200	8	65		✓	✓		✓
	75	175	6	50		✓	✓		✓
Resists tarnishing. Building and decorative components	170	470	15	100	✓		✓		
For highly stressed components in corrosive environments where high wear and shock loads may be encountered. Pumps, bearings, non-sparking tools, bushings, housings	200	550	18	130	✓	✓	✓		✓
	220	550	20	120	✓	✓	✓		✓
	280	650	12	150	✓	✓	✓		✓
	380	750	5	185	✓	✓	✓		
Seawater handling components, propellers	275	630	18	150		✓			
High strength and corrosion resistance for the most arduous marine applications. Pipe fittings and flanges in chemical engineering	100	280	25	70		✓	✓		✓
	120	340	18	80		✓	✓		
	250	440	18	115	✓				
	230	440	18	115		✓			

Melting copper and copper-based alloys

The melting of copper and copper-based alloys presents special problems. Molten copper dissolves both oxygen and hydrogen and on solidification, the oxygen and hydrogen can combine to form water vapour which causes porosity in the casting, Figs 16.1–16.4. Without the presence of oxygen, hydrogen alone may also cause gas porosity. Alloys containing aluminium form oxide skins which can cause problems in castings. In other alloys, traces of aluminium can cause defects and residual aluminium must be removed. Special melting and metal treatment techniques have been developed to deal with these effects. These include fluxing, degassing and deoxidation treatments. Foseco supplies products for each of these treatments.

Foseco products for the melting and treatment of copper and its alloys

ALBRAL	Fluxes for treatment of alloys containing Al, they dissolve and remove alumina.
CUPREX	Oxidising fluxes for preventing hydrogen pick-up during melting, Table 16.3.

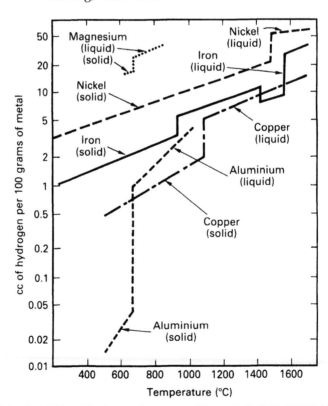

Figure 16.1 *Solubility of hydrogen in copper. (From Neff, D.V. (1989) Hydrogen and oxygen in copper,* AFS Trans., **97**, *439–450.)*

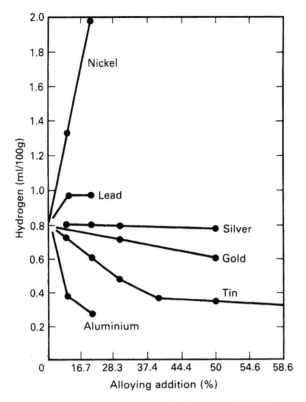

Figure 16.2 *Effect of alloying elements on hydrogen solubility in copper melts. (From Neff, D.V. loc. cit.)*

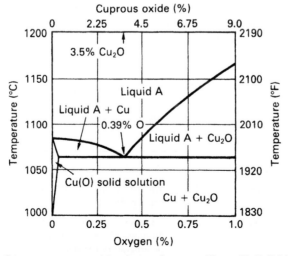

Figure 16.3 *Copper–copper oxide phase diagram. (From Neff, D.V. loc. cit.)*

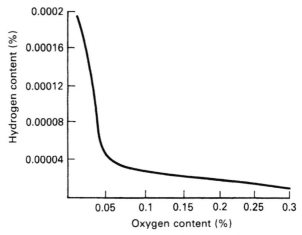

Figure 16.4   *Equilibrium between hydrogen and oxygen in copper melts. (From Neff, D.V.)*

CUPRIT	Neutral or reducing fluxes, they protect alloys from oxidation and reduce zinc loss, Table 16.4.
DEOXIDISING TUBES	For deoxidising copper and its alloys, Table 16.6.
ELIMINAL	Flux for removing aluminium from melts.
MDU	Mobile Degassing Unit for the removal of hydrogen.
LOGAS 50	Briquettes for the removal of hydrogen, Table 16.5.
PLUMBRAL	Covering and scavenging flux for treating high lead alloys.
RECUPEX	Fluxes for melting copper alloy swarf, skimmings and scrap.
RECUPEX 250	Reducing, protective flux for use when metal is held molten for a long time, e.g. during continuous casting.
SLAX 20	Slag coagulant.

Table 16.3   CUPREX oxidising fluxes and their applications

Product	Form	Application rate (%)	Alloys	Slag
CUPREX 1	Blocks	1	Commercial copper, gunmetal	Fluid
CUPREX 100	Powder	0.5–1	Tin/lead bronzes (<10% Pb) and copper–nickel alloys	Fluid
CUPREX 160	Powder	1–2	Commercial copper, bronzes, gunmetal, Ni–brass alloys melted in crucible or reverberatory furnaces	Plastic, dry

Table 16.4   CUPRIT reducing fluxes and their applications

Metal	Furnace	CUPRIT type	Recommended procedure
Brass Brazing metals Gilding metals	Crucible	1	Use 1% addition rate of CUPRIT. Place briquettes in the bottom of the hot crucible and charge metal on top. Leave the slag intact until the crucible is withdrawn from the furnace.
		81 49	Add 1% CUPRIT at an early stage in melting. Leave slag intact until the crucible is withdrawn.
	Small reverberatory	1	Place briquettes at the bottom of the hot furnace and add the charge. Use 1% CUPRIT.
	Electric	49	Add 0.5% CUPRIT in two stages, add the major portion to the heel and the remainder for final drossing-off. Skim before pouring.
HC copper	Crucible Electric	81	Add 1% CUPRIT at an early stage in melting. Leave the slag intact until the metal is tapped or the crucible withdrawn.
Brass Brazing metals Gilding metals Comm. copper	Electric low freq. induction	81	0.75% CUPRIT is needed. Add 0.6% together with charge, stir in the balance before drossing-off. More flux may be needed if the charge consists of swarf.
Brass Brazing metals Gilding metals Al–bronze Si–bronze Mn–bronze	All types of reverb. furnace	81	Add 0.5% at the beginning of melting. During melting add more to maintain a flux cover. 1% total may suffice depending on the surface of molten metal exposed.

Table 16.5   LOGAS 50 degassing tablets

Unit No.	1A	3B
Melt wt. treated kg	0–50	250–380

Table 16.6   Grades of DEOXIDISING TUBES and their use

Alloy	DEOX. TUBE	Weight of melt (kg)					
		25	50	75	100	200	400
Commercial copper	DS	2 × DS3	3 × DS4	6 × DS3	6 × DS4	3 × DS6	6 × DS6
HCC (high conduct.)	DS & CB	1 × DS1 + 1 × CB3	1 × DS2 + 2 × CB3	1 × DS3 + 3 × CB3	2 × DS4 + 1 × CB6	2 × DS4 + 2 × CB6	1 × DS6 + 4 × CB6
Brass	DS	1 × DS1	1 × DS2	1 × DS3	1 × DS4	2 × DS4	1 × DS6
Bronze & gunmetal	DS	1 × DS2	1 × DS3	1 × DS4	1 × DS5	3 × DS4	4 × DS5
Al-bronze & Mn–bronze	E	1 × E1	2 × E1	3 × E1	2 × E3	4 × E3	8 × E3
Nickel–silver castings	E & DS	1 × E3 + 1 × DS2	2 × E3 + 1 × DS4	3 × E3 + 2 × DS3	4 × E3 + 2 × DS6	8 × E3 + 1 × DS6	16 × E3 + 2 × DS6
Nickel–silver for hot & cold working	NS	1 × NS4	2 × NS4	3 × NS4	1 × NS6	2 × NS6	4 × NS6
Ni–bronze Cu–Ni alloys	MG	2 × MG5	3 × MG5	2 × MG6	3 × MG6	6 × MG6	12MG6

CUPREX oxidising fluxes and their applications

CUPREX formulations evolve oxygen to produce oxidising conditions and a scavenging gas to remove most of the dissolved hydrogen, thus preventing the steam reaction which causes porosity in castings. CUPREX also forms a flux cover to prevent the pick-up of more hydrogen from the furnace atmosphere and remove non-metallic material, Table 16.3.

CUPRIT neutral or reducing fluxes

The CUPRIT range is produced in briquette and powder form:

Briquettes   CUPRIT 1
Powder   CUPRIT 49, 81, 103

CUPRIT briquettes are pre-weighed while the powders are available in pre-weighed packets or in bulk. The main functions of CUPRIT are:

To form a protective blanket over the metal during melting to prevent contamination of the melt from the furnace atmosphere and to protect alloying elements, especially zinc, from oxidation, thereby suppressing zinc fume and the formation of showers of zinc oxide in the air.

To dissolve impurities from the melt.

To form an inert atmosphere for the melting of high conductivity copper (CUPRIT 81 flux).

To provide a mould and launder cover for the direct-chill casting of brass and copper (CUPRIT 103 flux).

Rotary degassing of copper and its alloys

The Mobile Degassing Unit (Fig. 6.2) is effective for removing hydrogen from copper melts and should be used in the way described for aluminium alloys in Chapter 6.

LOGAS degassing units

LOGAS degassing units comprise a mixture of chemicals which, on contact with molten metal, decompose to release a steady stream of non-reactive gas. LOGAS is carefully dried and packed in foil, so the gas bubbles contain very little hydrogen and are able to flush out hydrogen from the melt.

Deoxidants for copper and its alloys

The ideal deoxidant should function as follows:

1 It should combine with all the oxygen present to form a fluid slag.
2 Deoxidation products should not be entrained in the solidified casting.
3 Residual deoxidant should not adversely affect the physical properties of the alloy and should prevent further oxidation during pouring.

Phosphorus satisfies most of these requirements, but a residual content of 0.025% is necessary to ensure adequate deoxidation. This can seriously affect the conductivity of pure copper and causes embrittlement of high nickel bearing alloys.

Alternative deoxidants are:

MAGNESIUM: Very effective and it eliminates the harmful effects of sulphur, but the oxide formed tends to remain entrapped in the metal at grain boundaries, causing embrittlement.

MANGANESE: An excellent deoxidant, present in DEOXIDISING TUBES E. Manganese imparts some grain refinement.

CALCIUM: A good deoxidant, although metal fluidity is slightly reduced.

SILICON: Deoxidises well but the oxide formed may affect the surface appearance and pressure tightness of the casting.

BORON: A satisfactory deoxidant having some grain-refining action. Excess can cause embrittlement.

DEOXIDISING TUBES L are also available for commercial and h.c. copper, Ni–bronze, Cu–Ni alloys and Al–bronze. They contain lithium and remove hydrogen as well as deoxidise.

Copper-based alloy castings are usually made from charges using pre-alloyed ingot together with foundry returns (runners, risers and scrap castings). Such internal scrap must be carefully segregated to avoid mixing of metal of different specifications. With successive remelting there will be a tendency to lose volatile elements, particularly zinc, and to pick up contaminants such as iron. The level of residual phosphorus may vary, depending on the deoxidation practice used, and it must be carefully monitored.

The alloys are frequently melted in gas-fired furnaces, usually crucible furnaces. Medium frequency induction furnaces are also used with silica or alumina linings. Clay–graphite or silicon carbide crucibles can also be used, the electrical conductivity of the crucible allowing it to absorb induction power, yielding higher crucible temperatures and reduced stirring in the melt.

The melting and treatment of each of the main alloy types will be dealt with in turn.

Melting and treatment of high conductivity copper

The quality of high conductivity copper is measured by its electrical conductivity. Pure copper in the annealed condition has a specific electrical resistance of 1.72 microhms per cubic cm at 20°C. This is said to have 100% electrical conductivity IACS (International Annealed Copper Standard units). Cast copper can have a conductivity of 90% IACS and has both electrical and thermal applications since high electrical conductivity also implies high thermal conductivity. Many of the impurities likely to be present in copper lower its electrical conductivity seriously, Table 16.7.

Cu–C (HCC1) copper is used for water-cooled tuyeres and electrode clamps, it must have 86% IACS minimum so must be of high purity with only small additions of Cr or Ag to extend the freezing range and make casting easier.

For less onerous duties, copper having tin or zinc up to 2% may be used. A degree of conductivity is sacrificed to allow better casting properties and for ease of machining.

Where greater hardness and strength are required, copper–chromium castings CC1-TF may be used. This alloy requires to be heat treated (1 hour at 900°C, followed by quenching to room temperature and reheating to 500°C for 1–5 hrs) to realise its full properties.

High purity copper is particularly prone to gas porosity problems due both to hydrogen and the hydrogen/oxygen reaction which occurs if any

Table 16.7   The effect of impurities and alloying elements on the
electrical conductivity of pure copper

Impurity	%	% IACS
Aluminium	0.1	85
Antimony	0.1	90
Arsenic	0.1	75
Beryllium	0.1	85
Cadmium	1.0	90
Chromium	1.0	80
Calcium	0.1	98
Iron	0.1	70
Magnesium	0.1	94
Manganese	0.1	88
Nickel	0.1	95
Phosphorus	0.1	50
Silicon	0.1	65
Silver	1.0	97
Tin	1.0	55
Zinc	1.0	90

oxygen is present in the molten metal. Steps must be taken, during melting, to exclude both hydrogen and oxygen from the melt. The principles involved are:

Melt quickly, using the lowest temperature possible, under a reducing cover flux
Purge with an inert gas to remove hydrogen
Add deoxidants to remove residual oxygen, ensuring that residual deoxidant does not reduce the conductivity

Melting

The charge materials must be carefully selected to avoid impurities which can reduce the conductivity. Before charging, the copper must be clean and degreased to avoid any hydrogen-containing contaminants. Clean and dry crucibles, lids, plungers and slag stoppers must be used. The crucible should be preheated before charging to minimise the time that the copper is solid and unprotected by flux. Melt down under a reducing cover of CUPRIT 81; the flux should be placed in the bottom of the crucible prior to charging. 1 kg of CUPRIT 81 is needed per 100 kg of metal.

Degassing

Hydrogen is removed from the melt by bubbling an inert gas through the melt. This can be done using argon or nitrogen using the Mobile Degassing Unit (see Chapter 6) or less effectively by injecting gas through a graphite tube immersed deep into the melt. 50–70 litres of gas are needed for each 100 kg of copper.

An alternative way to degas is to plunge LOGAS 50 briquettes into the melt. LOGAS is a granular material, strongly bonded and formed into a weighed unit with high surface area/volume ratio to ensure maximum contact area with the liquid metal. On contact with the metal, LOGAS 50 decomposes releasing a steady stream of non-reactive gas which flushes out the hydrogen. LOGAS 50 units are packed in foil, they are of annular shape having a central hole into which a refractory-coated steel plunger can be inserted, Table 16.5.

Treatment takes from 3 to 10 minutes depending on the size of the melt. Some loss of temperature occurs during treatment, so the initial treatment temperature must be chosen accordingly. The minimum temperature practicable should be used.

Deoxidation

A number of deoxidants are available for copper (Table 16.6). They combine with the dissolved oxygen in the metal forming stable oxides which float out of the melt. Phosphorus is the most widely used deoxidant for copper and its alloys because of its effectiveness and low cost. It must be used sparingly with high conductivity copper since any residual phosphorus left in solution seriously lowers the conductivity of the copper (Table 16.7).

The recommended practice is to use phosphorus to remove most of the dissolved oxygen and to complete the deoxidation with a calcium boride or lithium-based deoxidant.

The precise quantity of deoxidant needed depends on the melting practice used. Simple tests can be made in the foundry to observe the solidification characteristics of the melt. Open-topped cylindrical test moulds having impressions about 75 mm high by 50 mm diameter are needed. They can be formed in a cold-setting resin or silicate sand mixture. When the melt is ready for deoxidation, a sample of the copper should be ladled into one of the moulds and allowed to solidify. If the head rises appreciably as shown in Fig. 16.5a very gassy metal is indicated. DEOXIDISING TUBES DS containing phosphorus must be plunged and further test castings made. At the point when the quantity of phosphorus added results in a shallow sink in the head, as in Fig. 16.5b, it can be assumed that the residual phosphorus content of the melt is nil and a small amount, about 0.008% of oxygen, remains.

Deoxidation is now completed by plunging DEOXIDISING TUBES CB or L, adding sufficient to produce a test casting having a head with a deep sink

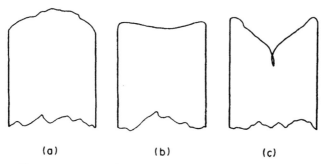

(a)   (b)   (c)

Figure 16.5 *The appearance of test castings: (a) Gassy metal. (b) Partially deoxidised. (c) Fully deoxidised.*

as in Fig. 16.5 c. The melt is now in a condition to produce castings free from porosity. The approximate additions needed are shown below:

Weight of melt

DEOXIDISING TUBES	25 kg	50 kg	75 kg	100 kg	200 kg	400 kg
DS & CB	1 × DS1 1 × CB3	1 × DS2 2 × CB3	1 × DS3 3 × CB3	2 × DS4 1 × CB6	2 × DS4 2 × CB6	1 × DS6 4 × CB6

DEOXIDISING TUBES L, containing lithium, can be used as the final deoxidant in place of DEOXIDISING TUBES CB. An application rate of 0.018–0.02% of product should be used. In addition to being an excellent deoxidant, lithium also removes traces of hydrogen. This is found to reduce the incidence of cracks in complex cast shapes.

Casting conditions

HC copper, being almost pure copper, has an extremely short freezing range. It is very weak at the point of solidification so that moulds and cores must not be too strong. Resin bonded sand is suitable and the resin percentage must be as low as possible, the minimum necessary for handling the mould and cores. Gating should be designed to minimise turbulence on pouring, in order to avoid the possibility of oxygen pick-up, Figs 16.6 16.7 (see Chapter 7). Feeding of the castings follows the practice used for steel castings (see Chapter 17).

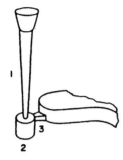

Figure 16.6   *Gating high conductivity or commercial copper castings, single ingate.*
The simplest form of gating for small to medium size castings.
1. Tapered sprue to reduce formation of air bubbles.
2. Deep basin to receive first turbulent impact of metal.
3. Ingate tapering out to reduce metal velocity. Note position of ingate at top of sprue basin.

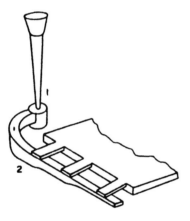

Figure 16.7   *Gating high conductivity or commercial copper castings, multiple ingates.*
For castings of large surface area where more than one ingate helps to fill mould uniformly.
1. Sprue and basin as in Fig. 5.6.
2. Progressively narrowing runner to keep runner bar full; this reduces dross formation. Note runner bar extention to trap dross.

Recommended casting temperatures

Light castings	< 15 mm section	1250°C
Medium castings	15–40 mm section	1200°C
Heavy castings	>40 mm section	1150°C

Melting and treatment of high conductivity copper alloys

Copper–silver

Silver additions should be made in the form of Cu–Ag master alloy and introduced into the melt after degassing but prior to deoxidation. The same dual deoxidation process used for pure copper is recommended.

Copper–cadmium

Degassing and deoxidation by the dual treatment must be completed before cadmium is added. The molten copper can be tapped directly onto pure cadmium metal as the metal is transferred from the melting furnace to a pouring ladle. The use of a Cu–Cd master alloy is preferable, since lower cadmium losses occur. Molten cadmium evolves toxic brown fumes so good ventilation is needed.

Copper–chromium

Cu–Cr master alloy should be added after degassing but before deoxidation. The chromium alloy should be thoroughly stirred in to ensure a homogeneous solution. A chromium loss of 10–30% may be expected depending on the state of oxidation of the melt. Phosphorus additions should only be made if a test casting shows a rising head. Normally the chromium addition and a final deoxidation with calcium-boride or lithium is sufficient. Any residual phosphorus left in the alloy will upset its response to heat treatment.

Commercial copper

Commercial copper castings contain up to 2% of tin and/or zinc for ease of casting and machining. The conductivity is reduced to a minimum of 55% IACS but this is adequate for many applications. A simpler fluxing and deoxidising technique can be used. Melting can be carried out under oxidising conditions and phosphorus alone can be used for deoxidation. Hydrogen degassing is not usually necessary since CUPREX oxidising fluxes evolve oxygen and scavenging gases which eliminate most of the hydrogen.

Treatment

Melt down under an oxidising cover of CUPREX (either CUPREX 1 blocks or CUPREX 100 powder), Table 16.3. Four 250 g blocks should be used for 100 kg of metal (1%). The CUPREX should be placed in the bottom of the empty crucible which is preheated to redness. In other furnaces, add the CUPREX at an early stage in melting. The fluid slag must be removed before deoxidation, using SLAX to thicken the slag if necessary. Deoxidise before pouring using DEOXIDISING TUBES DS at the following rate:

Wt. of melt (kg)	25	50	75	100	200	400
No. of tubes	2 × DS3	3 × DS4	6 × DS3	6 × DS4	3 × DS6	6 × DS6

Casting conditions

See recommendations for HC copper.

Recommended casting temperatures

Light castings	<15 mm section	1250°C
Medium castings	15–40 mm section	1200°C
Heavy castings	>40 mm section	1150°C

Melting and treatment of brasses, copper–zinc alloys

Effect of added elements

Aluminium: Unless added for a definite purpose, as in diecasting brass, it should be absent. It improves fluidity and definition, which is valuable in diecastings, but it oxidises readily causing oxide films and inclusions which may cause porosity and unsoundness in sand castings.

Iron: Small quantities have a grain-refining effect and increase hardness and tensile strength.

Lead: Improves machinability. Lead is insoluble in brass and exists as globules, which should be dispersed as uniformly as possible. It must not be allowed to segregate.

Manganese: Sometimes used as a deoxidant, its effect is similar to iron.

Nickel: Improves the mechanical properties and increases corrosion resistance. It also has a tendency towards grain refinement.

Phosphorus: Combines with any iron present, increasing hardness. Reduces grain growth and improves fluidity.

Silicon: Makes founding more difficult but improves corrosion resistance and fluidity.

Tin: Raises tensile strength and hardness at the expense of ductility and improves corrosion resistance and fluidity.

Principles of melting and treating brasses

Zinc vapour pressure in molten brass is sufficient to prevent ingress of hydrogen into the metal, so a neutral or reducing atmosphere is not deleterious. An oxidising atmosphere must be avoided since it would cause loss of zinc though oxidation. Zinc can also be lost through volatilisation, so a liquid flux cover is needed. To avoid zinc loss, the charge should be melted as quickly as possible and not be allowed to overheat.

Removal of impurities

ELIMINAL is a powdered flux range designed to reduce aluminium (and silicon) in copper-based alloys. Up to 0.5% Al may be removed from brass by means of ELIMINAL 2. Where higher levels exist, it is recommended that the charge is diluted with Al-free material to bring the Al content down to 0.5% or less. If the Al content is around 0.5%, the charge should be melted down under a cover of 0.5% by weight of ELIMINAL. This will also protect the zinc content of the melt. Before pouring, the metal should be brought to a temperature slightly higher than that required normally and 0.5% of ELIMINAL should be rabbled in or plunged to ensure maximum mixing, which is essential for efficient removal. The treatment is repeated until the Al content is reduced to the desired level. The metal is deoxidised immediately before pouring.

> When melts contain 0.4%Al, ELIMINAL removes about 40% of its own weight of Al
> When melts contain 0.2%Al, ELIMINAL removes about 25% of its own weight of Al
> When melts contain 0.1%Al, ELIMINAL removes about 20% of its own weight of Al

When aluminium has been removed to a low level, ELIMINAL will then remove silicon and manganese from copper alloys but at a slower rate than aluminium.

Melting brasses for sand castings

1 Heat up the crucible or furnace.
2 Place in the bottom of crucible or furnace CUPRIT 1 blocks equal to 1 kg per 100 kg of metal to be melted (Table 16.4).
3 Charge ingots and scrap and melt down as rapidly as possible, maintaining an intact cover of fused flux. CUPRIT 49 powder may be used instead of blocks. This should be added at the same rate as soon as the first part of the charge reaches a pasty condition.
4 Bring the metal up to pouring temperature and avoid overheating.
5 Immediately prior to pouring, plunge DEOXIDISING TUBES DS at the rate of one DS2 tube per 50 kg of metal and hold immersed for a few seconds. (The plunging tool must be preheated and coated with FIRIT or HOLCOTE 110 to protect the plunger and prevent contamination.)
6 When the metal is at the correct pouring temperature SLAX 20 may be added to reduce "flaring", the surface slag should be held back and the metal poured from underneath it. This will reduce "flaring" to the minimum.

CUPRIT blocks are recommended for use in reverberatory and similar hearth furnaces, since powder fluxes can be carried away by the draught from the burners.

Casting conditions (sand castings)

Brasses may be cast easily in green sand or chemically bonded sand moulds. Pinhole porosity may be a problem, often revealed when the casting is polished. It occurs particularly at higher pouring temperatures and can be avoided by application of a graphite-containing coating, such as ISOMOL 185, to the moulds and cores.

Running, gating and feeding

The running of brass castings does not present any real problem. Excessive turbulence in the mould should be avoided. Methods best suited to long freezing range alloys should be used (see Chapter 7), with unpressurised or only slightly pressurised systems based on ratios such as 1:4:6 or 1:4:4. This type of sprue/runner/ingate system can provide a useful source of feed metal to the casting as long as the gate remains unfrozen. Indeed, many thousands of castings (shell mouldings in particular) such as taps, valves, cocks etc. are made in this way without any supplementary form of feeding. The alloys go through a mushy stage during freezing and thin sections, below 10 mm, will often form a dense skin, free from porosity, while the centre of the section displays dispersed shrinkage porosity. So the castings may be pressure-tight throughout.

Recommended casting temperatures

	<15 mm	*15–40 mm*	*>40 mm*
60/40–65/35 alloys	1100°C	1050°C	1020°C
80/20–70/30 alloys	1150	1100	1070

Melting diecasting brasses

The diecasting brasses CuZn38Al-C (DCB1), CuZn39Pb1Al-C and CuZn29-AlB-C (DCB3) are cast by the gravity die (permanent mould) technique. The alloys contain aluminium which oxidises during melting forming a skin of oxide which results in sluggish pouring, so it is necessary to melt under a special flux such as the ALBRAL range. ALBRAL fluxes contain chemicals which dissolve alumina, removing it from the metal by flotation. The surface layer formed may be either a dry dross or a liquid slag, depending on the grade of ALBRAL used. From an efficiency aspect, a liquid slag performs best, but there may be difficulties in removing it in some operations. The following table indicates the types of ALBRAL available and their method of application.

ALBRAL fluxes for removing alumina

Product	Furnace type	Dross type	Addition during melting	Addition before pouring
ALBRAL 2	Crucible, reverb.	Fluid	Up to 1% to form a cover	0.25–0.5% plunge and rabble
ALBRAL 3	Bale out, induction	Dry	ditto	ditto

For melting diecasting brasses:

1 Preheat the crucible or furnace.
2 Charge ingots and scrap and commence melting.
3 When part of the charge becomes pasty, sprinkle ALBRAL 3 (1 kg for 100 kg of metal) over the surface.
4 Continue charging and melt under the protective cover as rapidly as possible.
5 When the metal is at pouring temperature, add a further quantity of ALBRAL 3 (0.5 kg for 100 kg of metal) and with a perforated saucer plunger, plunge the flux to the bottom of the melt, then with a rotary movement of the plunger, "wash" the flux in, bringing it into intimate contact with all parts of the melt in order to cleanse it of alumina particles.
6 After 2–3 minutes, withdraw the plunger and allow the melt to settle.
7 Deoxidise with DEOXIDISING TUBES DS (one DS 2 tube for 50 kg of metal).
8 When the metal is at the correct temperature (1100°C), ladle out as required, pushing the surface dross aside to leave a clean working area.
9 From time to time, after fresh metal has been added, skim away the dross and add fresh ALBRAL 3, washing in as before. Similarly, DEOXIDISING TUBES DS should be plunged occasionally to maintain maximum fluidity.

Recommended casting temperatures

Pouring temperatures range from 1150 to 1200°C and the die should be of heat-resisting cast iron or steel, die lives of 500–50 000 pours may be expected, depending on the complexity of the casting.

The dies should be provided with venting grooves to allow the displacement of air as the metal enters and overflow cavities to permit a flow-through of metal.

The dies should be sprayed with a water-based DYCOTE 36 which serves to:

> Lubricate the die to facilitate ejection
> Cool the die
> Prevent welding between the casting and the die
> Provide an insulating coating which promotes the running of the casting

The dies should be preheated to 350–400°C before casting to limit the thermal shock and to avoid misruns.

Pressure diecasting

Alloys CuZn39Pb1Al-C, CuZn29AIB-C (DCB3) and silicon brass, CuZn33Pb2Si-C, are the most frequently cast by pressure diecasting, since they have low melting points, minimising the deterioration of the die. Dies operate at a temperature of around 500°C, higher than in aluminium diecasting. Less die cooling is needed than for aluminium because heat losses are greater at the higher temperature.

Dissolved gas is not normally a problem in diecasting, hence degassing is not normally applied, but the melt must be kept clean from oxides, dross and inclusions.

Melting bronzes and gunmetals

Bronzes and gunmetals may be melted in crucibles, reverberatory furnaces or induction furnaces. Hydrogen is the main source of porosity problems, it may be derived from the products of combustion of the furnace gases, from water vapour in the atmosphere and from water in refractories and on scrap metal. Hydrogen is less soluble in copper alloys than in pure copper (Fig. 16.2) but it can cause severe porosity in castings due to the steam reaction with dissolved oxygen in the melt during solidification. For this reason it is often advisable to both degas and deoxidise bronze or gunmetal melts before casting. The technique used is the oxidation–deoxidation melting process. Oxidising conditions are maintained during melting, to minimise the hydrogen pick-up. The metal is degassed either by Rotary Degassing with nitrogen or by plunging LOGAS 50 briquettes, as described in Chapter 6, then it is deoxidised using phosphorus in the form of DEOXIDISING TUBES DS before casting.

Aluminium is a common and very deleterious impurity in sand cast gunmetals and bronzes. As little as 0.01% is enough to cause leakage of castings, because aluminium oxide films and stringers become trapped in the solidifying casting. ELIMINAL can be used to remove aluminium from the molten alloy (see page 245).

Melting for sand castings

1 Preheat the crucible or furnace.
2 Place CUPREX 1 blocks in the bottom of the hot crucible (1% of the charge weight).
3 Charge ingot and scrap.
4 Melt and bring to pouring temperature as quickly as possible. CUPREX evolves oxidising and scavenging gases which remove hydrogen, the flux cover protects the melt from further hydrogen absorption.
5 If extra degassing is needed, for example, if the charge materials are oily and dirty, degas by Rotary Degassing or plunge LOGAS 50 briquettes (see page 240).
6 Immediately before casting deoxidise the melt thoroughly by plunging DEOXIDISING TUBES DS at the rate shown in Table 16.6.
7 Check the correct pouring temperature, skim and cast without delay.

Recommended casting temperatures

Cu–Sn–Zn–Pb	*<15 mm*	*15–40 mm*	*>40 mm*
83/3/9/5	1180°C	1140°C	1100°C
85/5/5/5	1200	1150	1120
86/7/5/2	1200	1160	1120
88/10/2	1200	1170	1130
90/10 (P–bronze)	1120	1100	1030

Running and feeding

Methods best suited to long freezing-range alloys should be used, with unpressurised or only slightly pressurised systems based on ratios such as

1:4:6 or 1:4:4. This type of sprue/runner/ingate system can provide a useful source of feed metal to the casting as long as the gate remains unfrozen. Where additional feed is required, generous feeders are required on the heavier sections, as is usual for long freezing-range alloys (see Chapter 7).

Melting aluminium bronze

Melt down under a fluid cover of ALBRAL 2 to minimise oxidation and cleanse the melt of aluminium oxide films, then deoxidise.

Up to 1% of ALBRAL 2 is used to form the cover, then when the metal is at pouring temperature, plunge and rabble a further 0.25–0.5% of ALBRAL 2. After 2–3 minutes, leave the metal to settle then deoxidise by plunging DEOXIDISING TUBES E, at the rate given in Table 16.6. Deoxidation treatment coalesces suspended non-metallics and improves fluidity.

Recommended casting temperatures

Light castings under 13 mm section	1250°C
Medium 13–38 mm	1200°C
Heavy over 38 mm	1150°C

Melting manganese bronze

High Tensile Brass, CuZn35Mn2Al1Fe-C, CuZn25Al5Mn4Fe3-C (HTB1, HTB3)

Melt as for aluminium bronze (above). Finally deoxidise with DEOXIDISING TUBES E at the rate given in Table 16.6.

Recommended casting temperatures

Light castings under 13 mm section	1080°C
Medium 13–38 mm	1040°C
Heavy over 38 mm	1000°C

Melting high lead bronze

PLUMBRAL flux provides a cover during melting, which prevents oxidation losses. The plunging of PLUMBRAL before pouring scavenges the melt, removing impurities and assisting in the dispersion of the lead phase.

In crucible melting 0.5% by weight PLUMBRAL is added as soon as melting begins. A further 0.5% is added about 5 minutes before deoxidising the melt. Before pouring, the fluid slag formed is skimmed off, thickening it with dry silica sand if necessary.

In tilting, rotary or reverberatory furnaces, 0.5% of PLUMBRAL is added when the charge begins to melt. A further 0.5% is placed in the bottom of the preheated ladle, along with deoxidants, and the metal tapped onto them. Before pouring, the slag may be thickened with a coagulant to form a crust and the metal poured from beneath it.

Melting copper–nickel alloys

Nickel increases the solubility of hydrogen in copper melts, so it is necessary to melt under an oxidising cover of CUPREX, followed by degassing with the Rotary Degassing Unit or LOGAS 50 to eliminate the hydrogen, then finally to deoxidise.

For a 100 kg melt, use 2 kg of CUPREX 1 blocks, degas with one LOGAS 50 briquette and deoxidise with DEOXIDISING TUBES MG (3 MG6 tubes for 100 kg). Note that Cu–Ni alloys may be embrittled by phosphorus, so DEOXIDISING TUBES DS should not be used.

Recommended pouring temperatures

Light castings, under 15 mm section   1400°C
Medium castings, 15–40 mm section   1350°C
Heavy castings, over 40 mm section   1280°C

Filtration of copper-based alloys

Copper-based alloys, particularly those containing aluminium such as the Al–bronzes and some of the brasses, benefit greatly from filtration in the mould. SEDEX ceramic foam filters are recommended, see Chapter 8.

Chapter 17

Feeding systems

Introduction

During the cooling and solidification of most metals and alloys, there is a reduction in the metal volume known as shrinkage. Unless measures are taken which recognise this phenomenon, the solidified casting will exhibit gross shrinkage porosity which can make it unsuitable for the service for which it was designed. To some extent, grey and ductile cast irons are exceptions, because the graphite formed on solidification expands and can compensate for the metal shrinkage. However, even with these alloys, measures may need to be taken to avoid shrinkage porosity.

To avoid shrinkage porosity, it is necessary to ensure that there is a sufficient supply of additional molten metal, available as the casting is solidifying, to fill the cavities that would otherwise form. This is known as "feeding the casting" and the reservoir that supplies the feed metal is known as a feeder, feeder head or a riser. The feeder must be designed so that the feed metal is liquid at the time that it is needed, which means that the feeder must freeze later than the casting itself. The feeder must also contain sufficient volume of metal, liquid at the time it is required, to satisfy the shrinkage demands of the casting. Finally, since liquid metal from the feeder cannot reach for an indefinite distance into the casting, it follows that one feeder may only be capable of feeding part of the whole casting. The feeding distance must therefore be calculated to determine the number of feeders required to feed any given casting.

The application of the theory of heat transfer and solidification allows the calculation of minimum feeder dimensions for castings which ensures sound castings and maximum metal utilisation.

Natural feeders

Feeders moulded in the same material that forms the mould for the casting, usually sand, are known as natural feeders. As soon as the mould and feeder have been filled with molten metal, heat is lost through the feeder top and side surfaces and solidification of the feeder commences. A correctly dimensioned feeder in a sand mould has a characteristic solidification pattern: that for steel is shown in Fig. 17.1, the shrinkage cavity is in the form of a cone, the volume of which represents only about 14% of the original

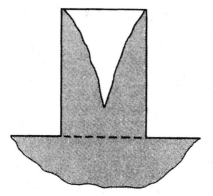

Figure 17.1 *Solidification pattern of a feeder for a steel casting.*

volume of the feeder, and some of this volume has been used to feed the feeder itself, so that in practice only about 10% of the original feeder volume is available to feed the casting. The remainder has to be removed from the casting as residual feeder metal and can only be used for remelting.

Aided feeders

If by the use of "feeding aids" the rate of heat loss from the feeder can be slowed down relative to the casting, then the solidification of the feeder will be delayed and the volume of feed metal available will be increased. The time by which solidification is delayed is a measure of the efficiency of the feeding aid. The shape of the characteristic, conical, feeder shrinkage cavity will also change and in the ideal case, where all the heat from the feeder is lost only to the casting, a flat feeder solidification pattern will be obtained, Fig. 17.2. As much as 76% of an aided feeder is available for feeding the casting compared with only 10% for a natural sand feeder. This increased

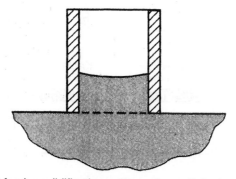

Figure 17.2 *Ideal feeder solidification pattern where all the heat from the feeder has been lost to the casting (schematic).*

efficiency means greatly reduced feeder dimensions with the following advantages for the foundry:

A greater weight of castings can be produced from a given weight of liquid metal
Smaller moulds can be used, saving on moulding sand binder costs
A reduction in the time needed to remove the feeder from the casting is possible
More castings can be fitted into the moulding box
Less metal melted to produce a given volume of castings
Maximum casting weight potential is increased

Feeding systems

Side wall feeding aids are used to line the walls of the feeder cavity and so reduce the heat loss into the moulding material. For optimum feeding performance, it is also necessary to use top surface feeding aids. These are normally supplied in powder form and are referred to as anti-piping or hot-topping compounds. Figure 17.3 illustrates how the use of side wall and top surface feeding aids extends the solidification times.

Calculating the number of feeders–feeding distance

A compact casting can usually be fed by a single feeder. In many castings of complex design the shape is easily divided into obvious natural zones for feeding, each centred on a heavy casting section separated from the

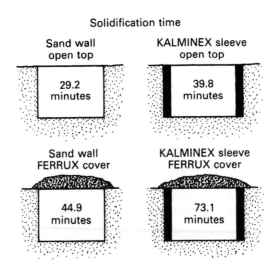

Figure 17.3 *Extension of solidification times with side wall and top surface feeding products for a steel cylinder 250 mm dia. and 200 mm high.*

remainder of the casting by more restricted members. Each individual casting section can then be fed by a separately calculated feeder and the casting shape becomes the main factor which determines the number of feeders required.

In the case of many extended castings, however, for example the rim of a gear wheel blank, the feeding range is the factor which limits the function of each feeder, and the distance that a feeder can feed, the "feeding distance", must be calculated.

The feeding distance from the outer edge of a feeder into a casting section consists of two components:

The end effect (E), produced by the rapid cooling caused by the presence of edges and corners

An effect (A), produced because the proximity of the feeder retards freezing of the adjacent part of the casting, Fig. 17.4

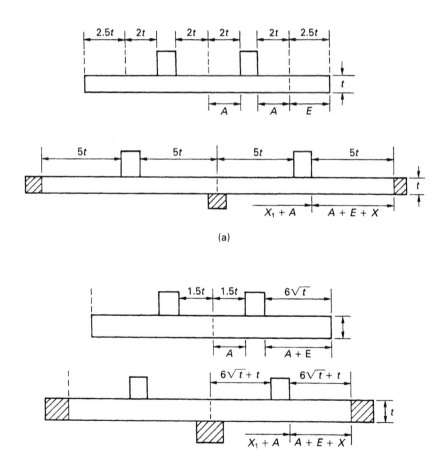

Figure 17.4 *Feeding distance in steel castings: (a) Plate (width:thickness >5:1) (b) Bar (width:thickness <5:1).*

Where a casting requires more than one feeder the distance between feeders is measured from the edge of the feeder, not from its centre; and when the feeder is surrounded by a feeder sleeve the distance between feeders is measured from the outside diameter of the sleeve.

The effective distance between feeders can be increased by locating a chill against the casting mid-way between the two feeders (X_1) and the natural end effect can be increased by locating a chill against the natural end (X). Chills should be of square or rectangular section with the thickness approximately half the thickness of the section being chilled.

There are therefore four possible situations:

Sections with natural end effect only $(A + E)$
Sections with natural end effect plus end chill $(A + E + X))$
Sections with no end effect (A)
Sections with no end effect plus chill $(A + X_1)$

Figure 17.4 shows the basis for calculating feeding distance in steel castings and all other ferrous alloys which freeze white, e.g. malleable and high alloy irons.

Ductile and grey iron castings

Alloy composition, casting section, mould materials and mould hardness all play a part in determining the actual feeding distance. The Tables 17.1 and 17.2 are guidelines for the specified conditions; variations from these conditions will result in other feeding distances;

Carbon equivalent 4.3%
Mould material   green sand
Mould hardness   90°B scale

Non-ferrous castings

Table 17.3 gives a feeding distance factor for some of the non-ferrous alloys; this factor is used in the calculation of an approximate feeding distance.

With reference to Fig. 17.4, FD is used in the following manner:

Bars (width:thickness <5:1)	*Plates (width:thickness >5:1)*
$A - 1.5t \times FD$	$2t \times FD$
$E = (6\sqrt{t} - 1.5t) \times FD$	$2.5t \times FD$
$X = t \times FD$	$0.5t \times FD$
$X_1 = (6\sqrt{t} - 0.5t) \times FD$	$3t \times FD$

Table 17.1   Feeding distance for ductile iron castings

Section thickness t (cm)	Feeding distance (cm)							
	Bars (width = thickness)				Plates (width = 5 times thickness)			
	$A + E$	$A + E + X$	A	$A + X_1$	$A + E$	$A + E + X$	A	$A + X_1$
1	19	21	17	19	19	21	17	19
2	37	41	33	37	31	35	27	31
3	50	56	44	50	34	40	28	34
4	55	63	47	55	41	49	33	41
5	57	67	47	57	52	62	42	52
6	62	74	50	62	62	74	50	62
7	72	86	58	72	72	86	58	72
8	83	99	67	83	83	99	67	83
9	93	111	75	93	93	111	75	93
10	104	124	84	104	104	124	84	104
11	114	136	92	114	114	136	92	114
12	124	148	100	124	124	148	100	124
13	135	161	109	135	135	161	109	135
14	145	173	117	145	145	173	117	145
15	155	183	125	155	155	185	125	155

The calculation of feeder dimensions

The majority of foundrymen, even today, decide on feeder dimensions on the basis of experience; however, the application of calculation based on established theory and experimental data ensures the most efficient design of natural and aided feeders. In this section some guidance is given for calculating feeder dimensions from first principles, whereas on page 286 there is a description of the various aids such as tables, nomograms and computer programs developed by Foseco to make the determination of feeder dimensions much easier.

The modulus concept

Although this concept has some shortcomings it is, with the exception of computer programs, the most widely used, acceptable and accurate method for calculating feeder dimensions.

Table 17.2 Feeding distance for grey iron castings

Section thickness t (cm)	Feeding distance (cm)							
	Bars (width = thickness)				Plates (width = 5 times thickness)			
	$A + E$	$A + E + X$	A	$A + X_1$	$A + E$	$A + E + X$	A	$A + X_1$
1	27	30	24	27	27	30	24	27
2	53	76	47	53	44	50	39	44
3	71	80	63	71	49	57	40	49
4	79	90	67	79	59	70	47	59
5	81	96	67	81	74	89	60	74
6	89	106	71	89	89	106	71	89
7	103	123	83	103	103	123	83	103
8	119	141	96	119	119	141	96	119
9	133	159	107	133	133	159	107	133
10	149	177	120	149	149	177	120	149
11	163	194	131	163	163	194	131	163
12	177	211	143	177	177	211	143	177
13	193	230	156	193	193	230	156	193
14	207	247	167	207	207	247	167	207
15	221	261	179	221	221	264	179	221

Table 17.3 Feeding distance factor for non-ferrous alloys

Casting alloy	Feeding distance factor (FD)
Al (99.99%)	2.50
Al 4.5% Cu	1.50
Al 7% Si	1.50
Al 12% Si	2.50
Cu pure	2.00
Cu 30% Ni	0.50
Brass	1.25
Al bronze	1.25
Ni Al bronze	0.50
Sn bronze	0.75

The solidification time of a casting section is given by Chvorinov's rule. where

$$t_c = kV_c^2/A_c^2 = kM_c^2 \qquad (1)$$

where:

t_c is the solidification time of the casting section
V_c is the volume of the casting section
A_c is the surface area of the casting section actually in direct contact with the material of the mould
k is a constant which is governed by the units of measurement being used, the thermal characteristics of the mould material and the properties of the alloy being cast.

M_c is the ratio of the volume of the casting section to its cooling surface area and is known as the casting's Geometric Modulus. It is expressed in units of length:

$$M_c = V_c/A_c \qquad (2)$$

The modulus formulae for some common casting shapes are shown in Fig. 17.5.

Generally, foundrymen, for the purpose of feeder size determination, are not directly interested in the exact solidification time but only that the feeder solidifies over a longer time than the casting. Having calculated the modulus of the casting section, therefore, the modulus of the feeder is calculated as:

$$M_F = 1.2 \times M_c \qquad (3)$$

where M_F is the modulus of the feeder required to feed a casting section having a modulus of M_c. This equation applies to natural feeders for most alloys. For grey and ductile iron castings, because there is a graphite expansion phase during solidification, the safety factor of 1.2 is considerably reduced and safety factors of 0.6 and 0.8 respectively are common.

The modulus extension factor (MEF)

The object of using feeding aids is to slow down the rate of heat loss from the surface of the feeder. It is possible to calculate how the improved thermal properties of the feeding aid compared with sand can reduce the feeder size.

Shape	Dimensions	Modulus
(a) Cube	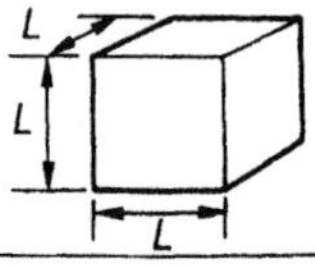 Side = L	$\dfrac{L}{6}$
(b) Cylinder	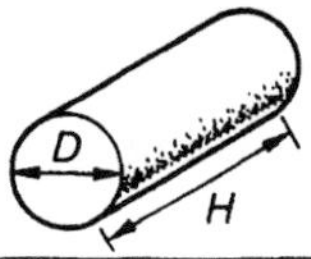 Diameter = D Height = H	$\dfrac{DH}{2D + 4H}$ Note: If H = D the modulus is $\dfrac{D}{6}$
(c) Disc	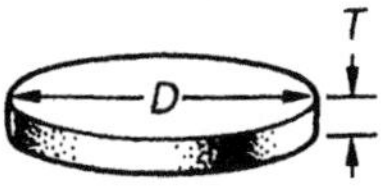 Diameter = D Thickness = T	$\dfrac{DT}{2D + 4T}$
(d) Bar or plate	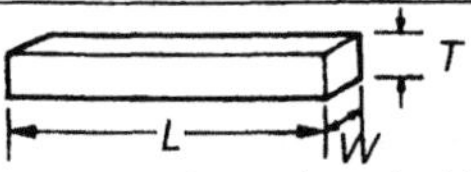 Length = L Width = W Thickness = T	$\dfrac{TWL}{2\,(TW + WL + LT)}$
(e) Endless cylinder (ends terminated by another part of casting)	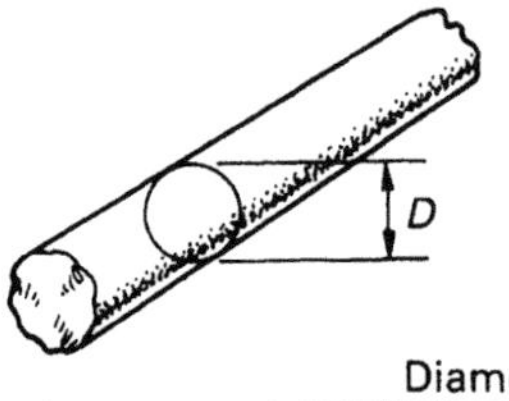 Diameter = D	$\dfrac{D}{4}$ Note: Because radial heat flow is faster than that from a flat surface, calculated moduli for endless cylinders may be reduced by multiplying by 0.85
(f) Endless plate (terminated on all sides by another part of casting)	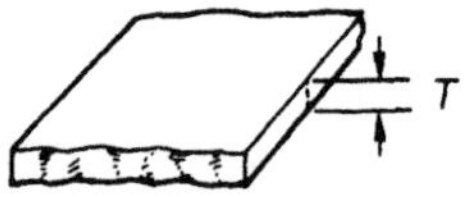 Thickness = T	$\dfrac{T}{2}$
(g) Endless bar (ends terminated by another part of casting)	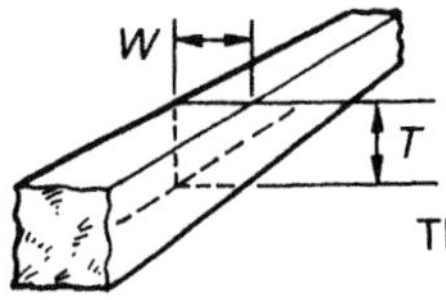 Thickness = T Width = W	$\dfrac{TW}{2\,(W + T)}$
(h) Endless hollow cylinder	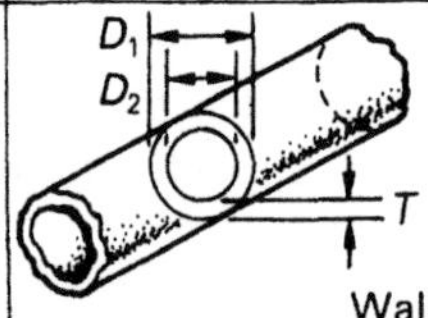 OD = D_1 Dia. core = D_2 Wall thickness = T	$\dfrac{D_1 - D_2}{4} = \dfrac{T}{2}$
(i) Annulus	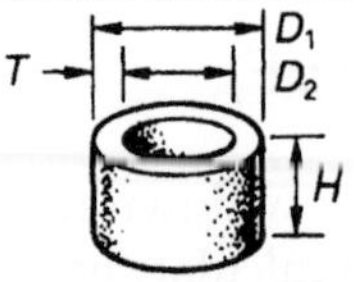 OD = D_1 Dia. core = D_2	$\dfrac{(D_1 - D_2)}{2\,(D_1 - D_2) + H}$ $= \dfrac{TH}{2\,(T + H)}$

Figure 17.5   *Modulus formulae for some common shapes.*

From equation (1) the solidification time for a feeder is expressed as:

$$t_F = kM_F^2 \qquad (4)$$

the constant k is composed of two parts:

The thermal characteristics of the mould material surrounding the whole of the feeder

The properties of the metal within the feeder

k can therefore be reduced to its two constituent components so that:

$$k = cf^2$$

where:

c is a constant depending on the properties of the metal being cast
f^2 is a constant depending on the properties of the mould material.

There is no significance in the fact that this constant is expressed as a square other than mathematical convenience.

Equation (4) can now be rewritten

$$t_F = C(fM_F)^2 \qquad (5)$$

The expression (fM_F) is known as the Apparent Modulus and f as the Modulus Extension Factor of the mould material surrounding all of the feeder's surfaces.

This approach provides a quantitative means of evaluating and comparing different feeding aids which should be designed to have the maximum possible Modulus Extension Factor compatible with the other properties necessary for a satisfactory product. In practice it is not customary to determine the absolute values of the constants relating solidification time to the modulus as these are seldom of interest.

Of greater concern is the improvement which can be expected from a variety of feeder lining materials when compared with the results obtained from the same size of feeder lined with the conventional moulding material – sand. For this purpose the Modulus Extension Factor (f) for sand is equated to unity and this serves as a basis for comparing other materials.

Example

Using this information it is possible to consider as an example a sound steel casting fed by one cylindrical feeder moulded in sand with a radius (r) of 16 cm and a height (h) of 32 cm. Because the feeder is attached to the steel casting, the bottom circular face of the feeder is a non-cooling face and the Geometric Modulus (M_F) of the feeder moulded in sand therefore is:

$$M_F \text{ (sand)} = \frac{\pi r^2 h}{2\pi rh + \pi r^2}$$

because in the example chosen $h = 32\,cm$; $r = 16\,cm$ and $h = 2r$:

$$M_F \text{ (sand)} = 2r/5 = 6.4\,cm$$

If in place of sand, a feeding aid system with a Modulus Extension Factor (f) of, for example, 1.6 were to be used to line the feeder cavity and to cover the top surface of the molten steel in the feeder, then the aided feeder would remain liquid for the same time as the sand-lined feeder if the Apparent Modulus of the aided feeder were to be equal to the Geometric Modulus of the sand-lined feeder, i.e. for the equal solidification times:

$$M_F \text{ (sand)} = f M_F \text{ (aided)} \tag{7}$$

so that the Geometric Modulus of the aided feeder would be:

$$M_F \text{ (aided)} = \frac{M_F \text{ (sand)}}{f}$$

or in the case chosen as the example:

$$M_F \text{ (aided)} = 6.4/1.6 = 4.0\,cm$$

a cylinder where $h = 2r$ having a Geometric Modulus of 4.0 cm would be 20 cm diameter × 20 cm high.

The use of the feeding aid system quoted in this example therefore represents a reduction of approximately 75% in the feeder volume needed to achieve the same solidification time. The values for Modulus Extension Factors vary widely according to the type of feeding aid, the size and shape of the feeder sleeve and even the shape of the casting being fed. Foseco publishes tables showing the dimensions of sleeves available together with the Geometric and approximate Apparent Modulus of each sleeve.

Determination of feeding requirements

Steel, malleable iron, white irons, light alloys and copper-based alloy castings

1. Calculate casting and feeder modulus

(a) Divide the casting into sections and determine the important volume-to-surface ratios according to Fig. 17.5:

$$\text{Modulus} = \frac{\text{Volume}}{\text{Cooling surface}}$$

$$M_c = V_c / A_c$$

(b) Determine the required feeder modulus (M_f) using the factor 1.2:

$$M_F = M_c \times 1.2$$

(c) Provisional determination of the feeder from the feeder sleeve tables published by Foseco companies.

2. Calculate the feed volume requirement

Feeders, whose dimensions obtained above satisfy modulus requirements, do not necessarily always satisfy the total feed metal demand of the casting section. This must always be checked and if the feeder is found to contain insufficient available feed metal the feeder dimensions must be increased. Generally it is preferable to retain the feeder diameter dictated by modulus considerations and increase the feeder height. Sometimes it is more convenient to increase the feeder diameter in place of or as well as height. Never reduce the feeder diameter below that necessary to meet modulus requirements. The following data is necessary:

(a) The proportion of feed metal available from the feeder which meets modulus requirements ($C\%$). Safe values although not necessarily the most efficient are:

> 33% if a Foseco sleeve is being used
> 16% if it is a live natural feeder (i.e. one through which metal has to flow before it reaches the casting cavity in the mould)
> 10–14% for other natural feeders

(b) The shrinkage of the alloy to be cast ($S\%$)
The shrinkage values for the principal casting alloys are given in Table 17.4.
(c) The weight of metal in the feeder under consideration (W_f).
The total weight of casting (W_c) which can be fed from a feeder of weight W_F is:

$$W_c = C/100 \times 100/S \times W_F \qquad (11)$$

If the total weight of casting section (W_T) which requires feeding is greater than W_C then increase the dimensions of the feeder until:

$$W_T = W_C$$

i.e. $$W_F = W_T \times 100/C \times S/100 \qquad (12)$$

Table 17.4   Shrinkage of principal
casting alloys

Casting alloy	Shrinkage (%)
Carbon steel	6.0
Alloyed steel	9.0
High alloy steel	10.0
Malleable iron	5.0
Al	8.0
AlCu4Ni2Mg	5.3
AlSi12	3.5
AlSi5Cu2Mg	4.2
AlSi9Mg	3.4
AlSi5Cu1	4.9
AlSi5Cu2	5.2
AlCu4	8.8
AlSi10	5.0
AlSi7NiMg	4.5
AlMg5Si	6.7
AlSi7Cu2Mg	6.5
AlCu5	6.0
AlMg1Si	4.7
AlZn5Mg	4.7
Cu (pure)	4.0
Brass	6.5
Bronze	7.5
Al Bronze	4.0
Sn Bronze	4.5

3. Calculate the dimensions of the feeder neck

(a) Top feeders
   No calculation of feeder neck dimensions is required. If possible, feeder
   sleeves should be used with breaker cores.
(b) Side feeders
   The required feeder neck dimensions are obtained from the calculation
   of the neck modulus (M_N) by applying the ratios:

$$M_C:M_N:M_F = 1.0:1.1:1.2$$

then using either the endless bar equation (Fig. 17.5) or the diagram in
Fig. 17.6

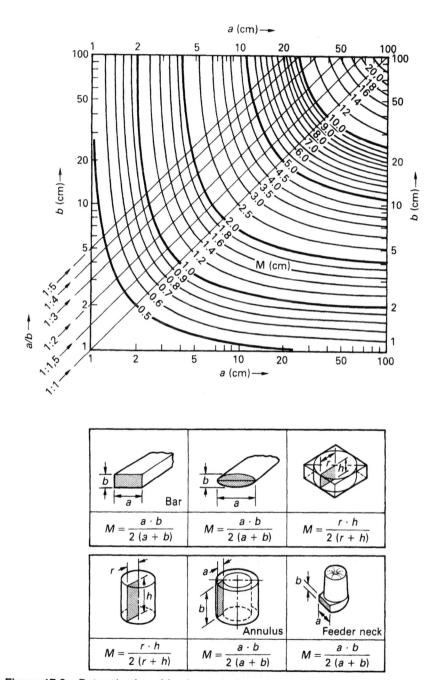

Figure 17.6 *Determination of feeder-neck dimensions.*

Grey and ductile irons

1. Calculate the casting modulus

Divide the casting into sections and determine the important volume-to-cooling surface area ratios according to Fig. 17.5:

$$\text{Modulus} = \frac{\text{Volume}}{\text{Cooling surface}}$$

$$M_C = V_C / A_C$$

2. Calculate the feeder modulus

(a) The graphite expansion which occurs during solidification of these alloys means that grey/ductile iron castings do not shrink for the full time during which liquid metal is present. The shrinking time (ST) is only a proportion of the total solidification time. This proportion, expressed as a percentage, is determined from the central and left-hand sides of Fig. 17.7, which is used in the following manner.

(b) Using the known carbon content, move parallel to the carbon line to the appropriate (Si + P) content at point A. Draw a line vertically until it intersects the casting modulus line at point B. Extend a line horizontally to the left until it intersects at point D with the line representing the estimated temperature of the iron in the mould. Read off shrinking time (ST) as a percentage of solidification time.

Effective feeder modulus is determined by:

$$M_F = M_C \times 1.2\sqrt{ST/100}$$

(c) Provisional determination of the feeder from feeder sleeve tables published by Foseco companies.

3. Calculate the feed volume requirement

Feeders, whose dimensions (obtained from the previous pages) satisfy modulus requirements, do not necessarily always satisfy the total feed metal demand of the casting section. This must always be checked and if the feeder is found to contain insufficient available feed metal the feeder dimensions must be increased. Generally it is preferable to retain the feeder diameter dictated by modulus considerations and increase the feeder height. Sometimes it is more convenient to increase the feeder diameter in place of or as well as height. Never reduce the feeder diameter below that necessary to meet modulus requirements.

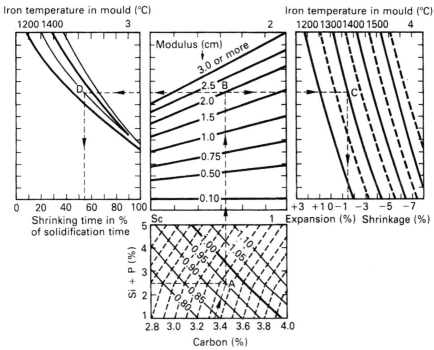

Estimation of shrinkage and shrinking time
from Analysis, casting modulus and metal
temperature in the mould.

Example: 3.35% C, 2.5% Si + P
Casting modulus:	2.0 cm
Casting temperature:	1300°C
Shrinkage:	1.6%
Shrinking time	55.0%

Figure 17.7 *Determination of shrinkage time and shrinkage for grey and ductile iron castings.*

The following data is necessary:

(a) The proportion of feed metal available from the feeder which meets modulus requirements (C%). Safe values although not necessarily the most effective are:

> 33% if a Foseco sleeve is being used
> 16% if it is a live natural feeder (i.e. one through which metal has to flow before it reaches the casting cavity in the mould)
> 10% for other natural feeders

(b) The shrinkage of the alloy to be cast. This is determined using the central and right-hand portion of Fig. 17.7. From point B draw a horizontal line until it meets the temperature of the metal in the mould (this must be estimated from the pouring temperature and is usually about 50°C less

than the pouring temperature). Read expansion or shrinkage (S) expressed as a percentage on the horizontal axis.
(c) Mould wall movement ($M\%$) depends on the hardness of the mould and can vary from zero for hard silicate or resin moulds to 2% for green sand moulds with a mould hardness of 85° (B scale). Add the mould wall movement from the expansion to give a final value of S. If this is positive or zero the casting will not need feeding.
(d) The weight of metal in the feeder under consideration (W_F). The total weight of casting (W_C) which can be fed from a feeder of weight W_F is

$$W_C = C/100 \times 100/S \times W_F \tag{11}$$

If the total weight of casting section (W_t) which requires feeding is greater than W_C then increase the dimensions of the feeder until:

$$W_T = W_C$$

i.e.
$$W_F = W_T \times 100/C \times S/100 \tag{12}$$

4. Calculate the dimensions of the feeder neck

(a) Top feeders
No calculation of feeder-neck dimensions are required. If possible feeder sleeves should be used with breaker cores.
(b) Side feeders
The required feeder-neck dimensions are obtained from the calculation of the neck modulus (M_N) by applying the ratios:

$$M_C:M_N:M_F = 1.0:1.1 \sqrt{ST/100}:1.2\sqrt{ST/100}$$

then using either the endless bar equation (Fig. 17.5 g) or the diagram in Fig. 17.6.

Foseco feeding systems

Introduction

Foseco provides complete feeding systems for foundries, comprising:

Sleeves – insulating and exothermic for all metals
Breaker cores – to aid removal of the feeder from the casting
Application technology – to suit the particular moulds and moulding machine used
Aids to the calculation of feeder requirements

Range of feeder products

The Foseco range of products comprises:

Product name	Type	Application casting alloy	Remarks
KALMIN S	Insulating	Al, Cu base iron, steel	extend solidification time by 2.0–2.2
KALMINEX 2000	Insulating and exothermic	iron, steel	extend solidification time by 2.5–2.7
FEEDEX HD	Highly exothermic	iron, steel	small riser-to-casting contact area use where application area is limited
KALMINEX	Insulating and exothermic	iron, steel	for large diameter feeders
KALBORD	Insulating	iron, steel copper base	in form of jointed mats for large feeders
KALPAD	Insulating	iron, steel copper base	prefabricated boards or shapes
KAPEX	Insulating	Al, Cu base iron, steel	prefabricated feeder lids, replace hot-topping compounds

KALMIN S, KALMINEX 2000, FEEDEX HD and KALMINEX products are supplied as prefabricated sleeves in a wide range of sizes and shapes, with or without breaker cores.

KALMIN S feeder sleeves

By using a high proportion of light refractory raw materials, a density of $0.45\,g/cm^3$ is achieved, ensuring highly insulating properties. KALMIN S

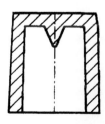

Table 1: Parallel conical insert feeder sleves

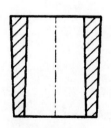

Table 2: Opposite conical feeder sleves

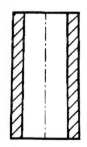

Table 3: Cylindrical feeder sleeves

Figure 17.8 *Foseco sleeve types.*

Figure 17.9   *Example of sleeve support on a moulding machine pattern plate.*

Figure 17.10   *An insertable KALMINEX sleeve with breaker core attached and a typical sleeve pattern.*

Figure 17.11   *The KALSERT system in operation. KALMINEX sleeves being inserted into green sand moulds on an automatic moulding line.*

sleeves are particularly suitable for aluminium, copper-base and iron alloy sand castings since the raw materials used are neutral to both the casting alloys and to the moulding sand. KALMIN S sleeves are supplied as "parallel conical", "opposite conical", or cylindrical sleeves for either ram-up application or insertion into cope or drag, Figs 17.8, 17.9, 17.10 and 17.11. They can be applied in the following moulding systems:

Moulding method	*KALMIN S feeder sleeve type*
Insert sleeves into the turned-over cope mould especially in automatic moulding lines	Parallel conical insert sleeves in conjunction with Foseco Sleeve Patterns
Insertion of the feeder sleeve from the top of the cope mould into the cavity created by means of a suitable sleeve pattern which is drawn after moulding	Opposite conical sleeve
Ram-up on the pattern plate in a machine moulding or a hand-moulding operation	Parallel conical insert feeder sleeve, opposite conical feeder sleeve or cylindrical sleeve
Insert sleeve in a vertically parted moulding system	Parallel conical insert and/or opposite conical feeder sleeve
Insert sleeves into shell moulds	Parallel conical insert and/or opposite conical feeder sleeve
Application of feeder sleeves in a drag mould over a feeder base or bridge core	Parallel conical feeder sleeve as floating sleeve system

KALMIN S feeder sleeves extend the solidification times by a factor of 2.0–2.2 compared to natural sand feeders of the same size. From these results, Modulus Extension Factors (MEFs) of 1.4–1.5 have been calculated. Though KALMIN S feeder sleeves can give more than 33% of their feeder volume to the solidifying casting, it is recommended that a maximum of one-third of the feed metal volume should be fed into the casting so that the residual feeder modulus is adequate in relation to the casting modulus at the end of solidification. For this reason, it is recommended to consider modulus as well as solidification shrinkage in order to determine the correct feeder. Foseco provides tables allowing KALMIN S feeders to be selected with the desired modulus, volume (capacity) and dimensions.

KALMINEX 2000 feeder sleeves

KALMINEX 2000 is a highly insulating and exothermic feeder sleeve material in the form of prefabricated sleeves for iron and steel casting in the modulus range between 1.0 and 3.2 cm. For light and other non-ferrous alloys the KALMINEX 2000 feeder sleeve material is not recommended. The manufacturing process specifically developed for this unique feeder sleeve material not only ensures a low density of 0.59 g/cm and, therefore, a high grade of insulation but also an additional exothermic reaction peaking at 1600°C. Due to the strength of the KALMINEX 2000 feeder sleeves, they can often be rammed up directly on the pattern plate on many moulding machines.

KALMINEX 2000 feeder sleeves can be applied in the following moulding systems:

Application method	KALMINEX 2000 feeder shape
Insert sleeves into the turned-over cope mould especially in automatic moulding lines	Parallel conical insert sleeves in conjunction with FOSECO Sleeve Patterns
Ram-up on the pattern plate in a machine moulding or a hand-moulding system	Parallel conical insert feeder sleeve, opposite conical feeder sleeve or cylindrical sleeve
Insert sleeve in a vertically parted moulding system	Parallel conical insert and/or opposite conical feeder sleeve
Insert sleeve into shell moulds	Parallel conical insert and/or opposite conical feeder sleeve
Application of feeder sleeves in a drag mould over a feeder base or on a bridge core	Parallel conical feeder sleeve as floating sleeve system

When determining the solidification times with KALMINEX 2000 feeder sleeves it has been found that they extend the solidification time by a factor of 2.5–2.7 compared to a natural sand feeder of the same size. From these results Modulus Extension Factors (MEFs) have been calculated between 1.58 and 1.64. Foseco provides tables allowing KALMINEX 2000 feeders to be selected with the desired modulus, volume (capacity) and dimensions.

Under specific practical conditions it has been found that the KALMINEX 2000 feeder sleeves can render 64% of their volume to the solidifying casting. When using feeders with the correct modulus it is necessary to take into account that the modulus of the residual feeder – if more than 33% of the feeder volume is fed into the casting – may not be adequate in relation to the casting modulus towards the end of the solidification. Therefore, it is essential to calculate shrinkage as well as modulus in order to determine the correct feeder sleeve.

FEEDEX HD V-sleeves

FEEDEX HD V feeder sleeves are used for iron and steel casting alloys. FEEDEX HD is a fast-igniting, highly exothermic and pressure-resistant feeder sleeve material. The sleeves possess a small feeder volume, a massive wall, but only a small riser-to-casting contact area, Fig. 17.12. They are, therefore, specially suited for use for "spot feeding" on casting sections which have a limited feeder sleeve application area. The sleeves are located onto the pattern plate using special locating pins, the majority are supplied with shell-moulded breaker cores. Owing to their small aperture, these breaker cores are not recommended for steel casting.

FEEDEX HD V-sleeves are particularly useful for ductile iron castings since with their low volume shrinkage of below 3%, a modulus-controlled KALMIN or KALMINEX 2000 feeder will often have more liquid metal than is necessary. The very high modulus and relatively low volume of FEEDEX HDV gives improved yield. In many ductile iron applications, the small breaker core aperture of the feeder means that the feeder is separated from the casting during the shakeout operation and the cleaning cost is reduced.

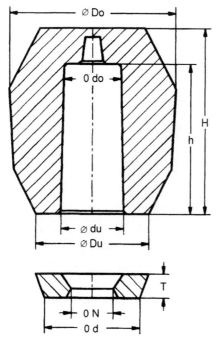

Figure 17.12 *FEEDEX HD V-sleeve.*

Figure 17.13   *The application of FEEDEX HD V-sleeves to ductile iron castings.*

When used in ductile iron applications, it is important to note that the high temperature reached in the highly exothermic feeder can cause residual magnesium in the iron to be oxidised so that there may be a danger of denodularisation on the casting–feeder interface. To avoid this, residual Mg should be greater than 0.045%, inoculation practice should be optimised and thick breaker cores used. Note that when calculating FEEDEX metal volume, only 50% of the capacity should be assumed since part of the metal in the feeder will be lamellar due to oxidation of the Mg in the feeder cavity.

Figure 17.13 shows examples of the use of FEEDEX HD V-sleeves on ductile iron castings.

KALMINEX feeder sleeves

KALMINEX exothermic-insulating feeder sleeves are used for all iron and steel casting alloys. They are supplied with feeder diameters from 80 to 850 mm for the modulus range between 2.4 and 22.0 cm and are suitable for larger-sized castings.

The manufacturing process specifically developed for this exothermic-insulating product and the selection of specific raw materials give a total closed pore volume of nearly 50%. The excellent heat insulation resulting from the low density (compared with moulding sand) is enhanced by an exothermic reaction.

When determining the solidification times with KALMINEX feeder sleeves it has been found that they extend the solidification time by a factor of 2.0–2.4 compared to the natural sand feeders of the same size. From these results Modulus Extension Factors (MEFs) of 1.4–1.55 have been found. Under practical conditions it has been found that KALMINEX feeders when adequately covered with KAPEX lids or a suitable APC (anti-piping compound) may render up to 64% of their contents into the casting. When using feeders with the correct modulus it is necessary to take into account that the modulus of the residual feeder – if more than 33% of the feeder volume is fed into the casting – may not be adequate in relation to the casting modulus towards the end of the solidification. Therefore it is essential to calculate shrinkage as well as modulus when determining the size of the feeder sleeves.

Foseco provides tables allowing KALMINEX feeders to be selected with the desired modulus, volume (capacity) and dimensions. Several different shapes of KALMINEX feeders are available, Fig. 17.14a,b,c. Breaker cores are generally made of chromite sand, although they can be produced in silica sand.

KALBORD insulating material

Although in theory there is no upper limit of inside diameter for using prefabricated feeder lining shapes for inside diameters above about 500 mm,

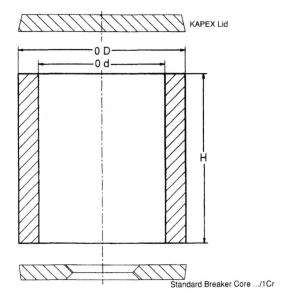

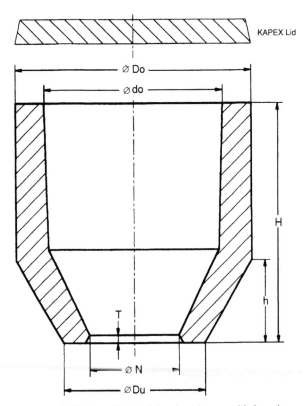

Figure 17.14   *(a) KALMINEX cylindrical feeder sleeve with breaker core and KAPEX lid. (b) KALMINEX TA sleeve.*

KAPEX Lid

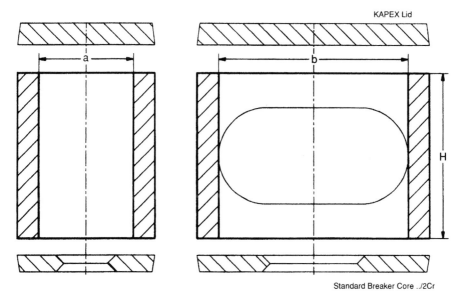

Standard Breaker Core ../2Cr

Figure 17.14 *(c) KALMINEX oval sleeve.*

manufacture, transport and storage become increasingly inconvenient. For this reason Foseco has developed KALBORD flexible insulating material in the form of jointed mats. They can be easily wrapped around a feeder pattern or made up into conventional sleeves as required for the production of insulating feeders for very large steel, iron and copper-based alloy castings, Fig. 17.15.

KALBORD mats are available with 30 mm and 60 mm thicknesses in widths up to 400 mm and lengths 1020 or 1570 mm. Their excellent flexibility permits the lining of irregular feeder shapes. The mat is most easily separated or shortened with a saw blade.

Produced from high heat insulating materials, 30 mm mats achieve a 1.3 fold and 60 mm mats a 1.4 fold extension of the modulus. It is recommended that KALBORD feeders are covered with FERRUX anti-piping powder.

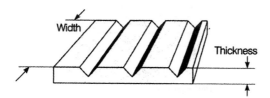

Width

Thickness

Figure 17.15 *KALBORD jointed mats.*

KALPAD prefabricated boards and shapes

KALPAD has been developed by Foseco to provide a lightweight, highly refractory insulating material to avoid metal padding and to promote directional solidification. If KALPAD insulating shapes are used the desired shape of the casting need not be altered. This increases yield and reduces fettling and machining costs. For this purpose KALPAD is used in copper-based metal and steel foundries and particularly in malleable iron and grey iron mass production.

Owing to a special manufacturing process and the use of alumina mineral fibres KALPAD shapes have a density of $0.45\,g/cm^3$ with more than 60% of the volume being closed pores which are the reason for the high insulation and refractoriness. During pouring KALPAD produces only negligible fumes and behaves neutrally towards moulding materials and casting metals.

When evaluating solidification times on KALPAD padded casting sections it has been found that they extend the solidification time by a factor of 2.25–2.5 compared with conventionally moulded castings. From these results Modulus Extension Factors (MEFs) of 1.5–1.58 have been calculated. It is recommended to use a factor of 1.5 if KALPAD shapes of 20–25 mm thickness are applied. The dimensions of KALPAD boards and shapes are shown in Fig. 17.16.

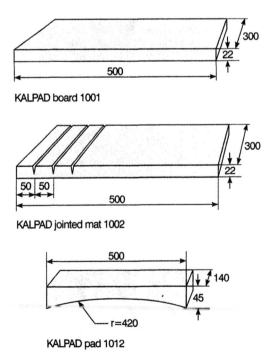

KALPAD board 1001

KALPAD jointed mat 1002

KALPAD pad 1012

Figure 17.16   *KALPAD prefabricated boards and shapes.*

KAPEX prefabricated feeder lids

KAPEX insulating feeder lids, Fig. 17.14a, are an improvement over the hot-topping powders in foundry use, being dust and fume free and giving repeatable feeding results. They can be applied to all feeders either exothermic, insulating or natural. The lids have an insulator density of $0.5\,g/cm^3$ and are purely insulating. Owing to their neutral behaviour towards moulding material and casting metal they are used in light metal and copper-based foundries as well as in high alloy steel foundries.

KAPEX KALMINEX 2000 lids are also available.

Breaker cores

Breaker cores for the reduction of the feeder-to-casting contact area enable feeders to be broken off or knocked off from many types of castings. In the case of very tough casting alloys where it is not possible to simply break off or knock off the feeder, the advantage of using breaker cores lies in the reduction of fettling and grinding costs for the removal of the feeder.

Besides the conventional types of breaker cores based on silica sand (Croning) and chromite sand, special breaker cores with a very small aperture are also in use in repetition iron foundries. These special breaker cores as shown in Table 17.5 are made from highly refractory ceramic.

Table 17.5 Application of breaker cores

Breaker core material	Casting metal	Feeder diameter (mm)
Silica sand	Steel	35–120
Silica sand	Grey iron, s.g. iron, non-ferrous metals	35–300
Ceramic	Grey iron, s.g. iron, non-ferrous metals	40–120
Chromite sand	Steel	80–500
Chromite sand	Grey iron, s.g. iron	200–500

Experience has shown that at least 70% of the breaker core area should be in contact with the casting, in order to level out the temperatures of the metal and the breaker core from the superheat upon or before reaching liquidus.

Some of the standard forms of breaker core available from Foseco are shown in Fig. 17.17. Foseco feeder sleeves can be ordered with or without breaker cores attached.

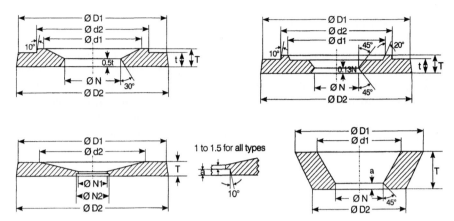

Figure 17.17 *Standard forms of breaker cores.*

The application of feeder sleeves

On large individual patterns

Sleeves of the correct dimensions are set on the individual pattern in the predetermined location and the mould is rammed around the sleeves. The base of the sleeve should not come into direct contact with the casting but be set on a sand step at least 10 mm thick or the sleeve should be fitted with a breaker core.

On pattern plates on moulding machines

If the pattern plate is accessible to the machine operator, the feeder sleeve is located by hand on the pattern plate. To avoid damage during machine moulding, sleeves should be supported by standing them on a pattern dummy or peg at the correct location and having the correct shape and height. Figure 17.9 shows one such arrangement.

Insert sleeves

Automatic moulding machines are capable of high output rates, machine operators are often no longer required, and in any case the pattern is no longer accessible. Foseco has recognised these changes and has developed insert sleeve application systems allowing fully automatic machine users to retain all the advantages of employing feeder sleeves without slowing down the moulding cycle.

A prefabricated feeder sleeve with strictly controlled dimensional tolerances is inserted into a cavity formed during the moulding operation by a sleeve pattern of precise dimensions located on the pattern plate, Figs 17.10 and 17.11.

The insert sleeve patterns are fixed by screwing them onto the casting pattern and they provide the cavity for the insert sleeve. Owing to the special sealing and wedging system no metal can penetrate behind the inserted sleeves and these cannot fall out from their seat during closing and handling of the mould.

The design of the insert patterns also forms highly insulating air chambers behind the inserted sleeves. This additional insulation increases the moduli of the insert sleeve feeders as follows:

FEEDEX insert sleeves	HDP	+5%
KALMINEX 2000 insert sleeves	ZP	+5%
KALMIN S insert sleeves	KSP	+4%

The insert sleeve patterns have a solid aluminium core with mounting thread and a highly wear-resistant resin profile. Insert sleeve patterns are available corresponding to the various types of insert sleeves.

Floating feeder sleeves

This is a relatively simple application technique with low feeder sleeve application cost since feeder sleeves are simply placed on the drag parting line. The method is applicable for all moulding machines having a horizontal mould parting line. No problems are encountered regarding

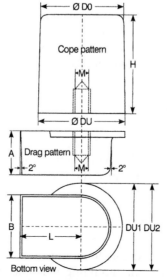

Figure 17.18 *Sleeve pattern for a floating sleeve.*

strength, springback etc. of the feeder sleeve. On high pressure moulding lines, cheaper and non-polluting insulating KALMIN sleeves can be applied.

A two-part sleeve pattern is used with an integrated feeder base and feeder neck, Fig. 17.18. The drag sleeve pattern is secured onto the drag pattern plate which creates a suitable location and positioning cavity for the corresponding feeder sleeve. The feeder sleeve is simply positioned on this location cavity, Fig. 17.19a. The cavity created by means of the cope sleeve pattern ensures location of the feeder sleeve while closing the mould. After pouring, the feeder sleeve floats along with the liquid metal, secures and seals itself tight into the mould wall cavity created by means of the cope sleeve pattern, Fig. 17.19b.

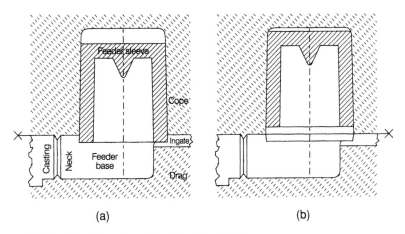

Figure 17.19   *Floating sleeve functional principle.*

The floating sleeve patterns incorporate maximum feeder-neck dimensions applicable to iron castings. For steel, light alloys and non-ferrous alloys, neck modulus can be modified to usual casting modulus equal to neck modulus. For full details, refer to Foseco leaflets.

Shell mould application

Sleeves may also be inserted into shell moulds. The principle is the same as for green sand moulding, special sleeve patterns are available which form ridges in the sleeve cavity which grip the inserted sleeve, Fig. 17.20.

DISA insert sleeve patterns

Insert sleeves can be applied equally to moulds with a vertical parting, such as those made on the Disamatic moulding machine. The sleeve pattern is divided – but off centre – one part being slightly smaller than the other. The

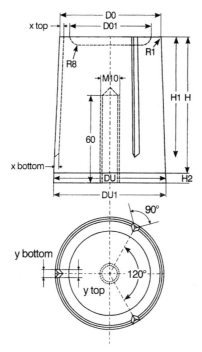

Figure 17.20 *Sleeve pattern for shell mould application.*

two parts are mounted on opposite sides of the Disamatic pattern plates with the sleeve located in the larger cavity and held in place by the exact vertical fit of the sleeve in the mould. When the mould is closed the second half holds the sleeve fully in position.

Application to cores

Feeder sleeves may also be inserted into cores. For example, ductile iron hubs are often fed by one or more side feeders located externally to the flange but the most efficient feeding method is by means of a sleeve located in the central core and connected to the casting at the point where the feed metal is really needed. The sleeve fits into the core and is held down by the cope when the mould is closed; the result is an improvement in yield, cleaning costs and casting soundness.

Williams Cores

The purpose of Williams Cores is to provide an aperture in the skin of the feeder so that the atmospheric pressure has access to the feed metal to

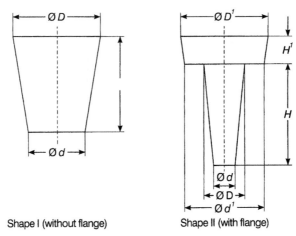

Shape I (without flange)   Shape II (with flange)

Figure 17.21   *Williams Cores.*

promote the feeding of the casting. Williams Cores are supplied in a range of sizes up to 66 mm diameter D, Fig. 17.21, in FEEDEX exothermic material. KALMIN S and KALMINEX 2000 parallel conical insert sleeves are manufactured with a Williams Wedge incorporated into the design, Fig. 17.8.

FERRUX anti-piping compounds for iron and steel castings

The FERRUX range includes anti-piping compounds of all types with reactions in contact with the molten metal which vary from very sensitive, highly exothermic to purely insulating. Described as examples are three grades of FERRUX manufactured in the UK which cover the requirements of the complete range of all ferrous alloys cast in all feeder diameter sizes.

All three grades have an exothermic reaction and one of them, FERRUX 707F, by expanding in use, incorporates the most modern technology. The examples detailed below therefore should only be considered as typical of the types of FERRUX grades and the technology which is available.

The anti-piping compound, pre-weighed and bagged, should be added in the bag to the surface of the metal immediately after pouring has been completed. It is advisable to design the feeder to pour slightly short so that a space can be left between the surface of the metal and the top of the mould. FERRUX will then be contained in this space. The recommended application rate is a layer which has a thickness equivalent to one-tenth of the diameter or 25 mm whichever is the greater. If after application the powder is not evenly distributed then the upper surface should be raked flat; normally this will not be found to be necessary.

FERRUX 16

This is a carbon-free, sensitive, fast reacting exothermic anti-piping compound of high heat output. After the exothermic reaction has ceased, a firm crust remains on top of the feeder. It is particularly recommended for use on feeders where rapid sculling takes place and where carbon contamination is to be avoided. Feeders where FERRUX 16 is most often employed are in the diameter range 25–200 mm.

FERRUX 101

This is an exothermic anti-piping compound of medium sensitivity. It is ideal for general steel foundry use on feeders of 150 mm diameter and upwards. It may also be used on iron casting feeders where the crust formed after the exothermic reaction has ceased, forms a good insulation against heat losses. The crust can be broken for topping up large castings. The absence of carbonaceous materials in the product ensures that no carbon contamination of the feed metal will occur.

FERRUX 707F

This is a medium sensitivity, exothermic anti-piping compound which expands during its reaction to approximately twice its original volume, to produce a residue of outstanding thermal insulation. In spite of the exothermic reaction, FERRUX 707F is virtually fume free and, in addition, because of the expansion and the product's lower density, the original weight of FERRUX 707F which has to be used for effective thermal insulation is usually only about half that of non-expanding grades. The low carbon content of this product will not normally affect metal quality in any significant way. FERRUX 707F is most generally employed for steel and iron feeders of 150 mm diameter and upwards.

Metal-producing top surface covers

THERMEXO is a powdered, exothermic feeding product which reacts on contact with the feeder metal to produce liquid iron at a temperature of about 2000°C. The product is designed for emergencies in case of metal shortage.

Even in the best foundries, occasionally the weight of metal left in the ladle is overestimated and a casting is poured short. The addition of a metal-producing compound may save the casting by providing the extra feed metal necessary. In such cases the foundry has nothing to lose by to try a metal-producing compound and it is for emergency reasons that every steel foundry should have a stock of THERMEXO.

FEEDOL anti-piping compounds for all non-ferrous alloys

The lower casting temperatures and the differing chemical requirements for non-ferrous alloys necessitate a completely different range of anti-piping compounds than that used on ferrous castings. FEEDOL is the name given to Foseco's range of anti-piping compounds for non-ferrous castings. As an example, two of the principal FEEDOL grades manufactured in the UK are described in detail below.

FEEDOL 1

This is a mildly exothermic mixture suitable for all grades of copper and copper alloys. The formulation does not contain aluminium and there is therefore no risk of contamination where aluminium is an undesirable impurity. After the exothermic reaction has ceased, FEEDOL 1 leaves a powdery residue through which further feeder metal can be poured if necessary. FEEDOL 1 is useful for feeders up to 200 mm diameter. For very large copper-based alloy castings such as, for example, marine propellers, FERRUX 707F is to be recommended.

FEEDOL 9

This is a very sensitive and strongly exothermic compound recommended for use with aluminium alloys. After the completion of the exothermic reactions the residue forms a rigid insulating crust. FEEDOL 9 is recommended for aluminium alloy feeders of all sizes.

Aids to the calculation of FEEDER requirements

Tables

Tables have been drawn up which will convert natural feeders to sleeved feeders for steel castings. No knowledge of methoding is required; all that is necessary to know are answers to the following questions:

(a) What are the dimensions of the natural feeder?
(b) What is the weight of the casting section being fed?
(c) What is the alloy composition?
(d) Is the casting with the natural feeder sound?

The tables will do the rest. The conversion, however, is very primitive for if the natural feeder is too large then the sleeved feeder will also be too large,

Table 17.6 Feeding guide for ductile iron castings

Weight of casting section (kg)	Sleeve type no. (weight)	Sleeve unit no. insert tapered
270.0	16/15 (19.5 kg)	KC3830
180.0	14/15 (13.9 kg)	KC3826
130.0	12/15 (9.8 kg)	KC3324
82.0	10/13 (5.7 kg)	KC3168
60.0	9/12 (4.8 kg)	KC3596
37.0	8/11 (3.0 kg)	KC3164
26.0	7/10 (2.2 kg)	KC3160
14.0	6/9 (1.4 kg)	KC3156
8.9	5/8 (0.92 kg)	KC3152
6.8	4/95 (0.77 kg)	KC3148
4.5	4/7 (0.55 kg)	KC3144
1.7	3.5/5 (0.23 kg)	KC3998

and conversely if the natural feeder is too small and causes shrinkage so will the sleeved feeder.

Similarly a simple table has been compiled for ductile iron castings. It is simple to use and requires no expert knowledge of methoding practice. Although in most cases the recommendation if followed will give a suitable feeder sleeve it is not necessarily the optimum size for a given casting section. The table compiled by Foseco in the UK is shown in Table 17.6.

Nomograms

A series of nomograms which relate the casting modulus, which has to be calculated, and the weight of the casting section to a suitable size of feeder sleeve has been developed. Two examples are shown in Fig. 17.22. Such nomograms have distinct disadvantages, they do not take into account many of the variables commonly found in steel foundries; they are, however, a significant step forward for feeder recommendations can be made without the need to know the original natural feeder dimensions.

FEEDERCALC

FEEDERCALC is a Foseco copyrighted PC computer program which enables the foundry engineer to make rapid, accurate calculations of casting weights, feeder sizes, feeder-neck dimensions and feeding distances and to

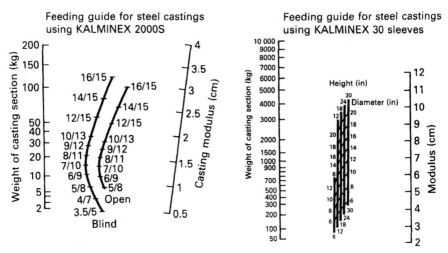

Figure 17.22   *Examples of nomograms used to determine suitable feeder sleeve dimensions.*

make cost analyses to quickly determine the most cost-effective feeding system for any given casting. The program is available in versions for iron castings and also for steel castings. A non-ferrous program is not available.

Calculating feeder sizes for aluminium alloy castings

Steel and ductile iron foundries use all the calculation methods available to determine the most effective feeders for their castings. FEEDERCALC is widely used and there is growing use of solidification modelling using computer programs such as SOLSTAR, MAGMASOFT etc. to simulate solidification with the casting in different positions before the casting is made.

Aluminium and other non-ferrous foundries do not usually use such calculation methods but instead rely to a large extent on the general principles described in Chapter 7 and experience to determine feeder size and position. The reason for this is mainly the difficulty of predicting the solidification pattern of long freezing range alloys, particularly if hydrogen gas is evolved as the casting freezes.

Index

Printed and bound by CPI Group (UK) Ltd, Croydon, CR0 4YY

08/05/2025

01864807-0001